To Leonard Susskind whose lectures and enthusiasm inspire me
Naveen Balaji U S

Preface

This book lays the mathematical foundations required to understand the majestic grandeur and essence of General Relativity (GR). My fascination with the ideas of the theory date back to high school when I gained substantial knowledge from books and Youtube videos. As an undergraduate, I wanted to get rid of the pop-science noise and formally study the theory. As all beginners do, I too faced difficulty understanding the mathematics which would often lead to me giving up pursuing further into the physics. Although there are books which lay strong mathematical foundations, they often overwhelm a beginner. I eventually found out that it was one's imaginative thinking that enables them to understand deeply and appreciate theories such as GR and using this approach I mastered the basics so much so that I could conduct two summer schools teaching GR to highly-motivated undergraduates. It is these schools that encouraged me to write a book, a book in which I can communicate deep ideas in a contemporary style.

The aim of this book is to present with precision, but as intuitively as possible, the foundations and main consequences of GR and it is written for students interested in exact mathematical formulations or indeed for anyone with a scientific mind irrespective of your educational field. The mathematical level of the seven chapters of the book is that of undergraduates of mathematics or physics. The book assumes the reader to possess a fair knowledge of Special Relativity and Electromagnetism and aims at communicating the concepts as intuitively as possible, constantly promoting the avant-garde and consciously avoiding the vicissitudes one faces in the conventionalist approach to physics and the labyrinthine choice of words in archaic texts.

I would like to thank Poojana Prasanna for the original and awe-inspiring illustrations, my friend Sujan Kumar S for accepting my invitation to join this project, and Vignesh T and Tarun SR for editing the book. I would also like to thank the students of my summer schools for providing honest reviews and Nishant Shetty for his deep insights.

Naveen Balaji U S
August 2018

Contents

1

Manifolds: A Pedestrian Approach

Imagine that your car's tyres record of all the information regarding where you had been, have been and will be going, all the events that had happened on the road is recorded and stored in each thread of the tyre, and since you are unsure of the events of the future, making assumptions that the tyres never wear out (no matter what happens to it and lasts forever) and that they are superelastic, we can safely assume that the tyres would possess an infinite number of threads. One can access the information stored in each thread to check for the type of terrain the tyre has travelled upon, and the shapes of the localized deformations it had experienced. If there are an infinite number of terrains, each unique (i.e. different from each other), upon which the tyres have travelled then there is an infinite number of deformation shapes (created on the tyres during travel). (see figure for a visual representation of this analogy)

Hence, we can satisfactorily say that all the deformation shapes, although unique, are after all nothing but mere closed geometric figures. Now, if we randomly pick two tyres and name them A and B, we can find the set of deformations they have experienced and since the terrains are unique at each point and an infinite number of events have occurred, chances are that the sets of deformations of tyre A around an arbitrary point, p, are completely different from those of tyre B, around an arbitrary point, q. The deeper picture here is that although the set of deformations are different, we can perform few operations (strictly mathematical) in order to make the deformations look similar. Hence, let us propose the following method-For every deformation in tyre A there is another in B which has a very close resemblance, at the same time, for every

Fig. 1.1. A visual representation of the analogy presented in this chapter. The bottom figure shows the different deformation shapes created on the tyre surface that are stored in the threads along with the information of the type of terrain in which the deformation occurred.

deformation in A there are multiple other deformations in B since all of them after all fall under the category of closed figures. The converse of the above statements is also true (i.e. from tyre B to A). Let's recap before concluding, we picked two tyres at random and named them A and B. We observed that not only are the sets deformations around arbitrary points on the tyres not the same but also that the deformations among the tyres had resemblances. At the end of the day, both the tyres had deformations which were nothing but closed shapes.

Alright then, let's conclude! A manifold is similar to our tyres, at localized points it represents Euclidean space just like how there were closed geometric deformations on localized points of the curved tyres, but ultimately it is nothing but a topological space because all deformations, although unique are nothing but closed shapes! More specifically a manifold is one in which at each point it has a neighbourhood which is *homeomorphic* to the *Euclidean space*. The deformations of the tyres could be made to look similar using the method I proposed, remember? When we say that for every deformation in tyre A there is another in B which has a very close resemblance, what it mathematically means is that every deformation in A has a *one-to-one* connection to another deformation in B. When we say that for every deformation in A there are multiple other, topologically similar deformations in B, it mathematically translates to stating that every deformation in A is connected onto multiple others in B. Since our deformations have both one-to-one and onto connections, mathematically we call them to be *bijective*. Do not forget that this bijective display of behaviour is limited only to the deformations in A, the converse is also true. In mathematics, there is a name assigned to the method we proposed- homeomorphism.

If you still crave for a more formal definition, here it goes: Suppose $f : A \to B$ is a bijective (one-to-one and onto) function between topological spaces A and B. Since f is bijective, the inverse f^{-1} exists. If both f^{-1} and f are continuous functions, then f is called a homeomorphism. Since the topological spaces A and B are homeomorphic, we denote them by the following: $A \cong B$.

Since we are comfortable with the meaning of a manifold, let us go a bit deeper. We observed that the set of all deformations around an arbitrary point p on tyre A was not similar to that around point q on tyre B, in other words, the sets of deformations around local-

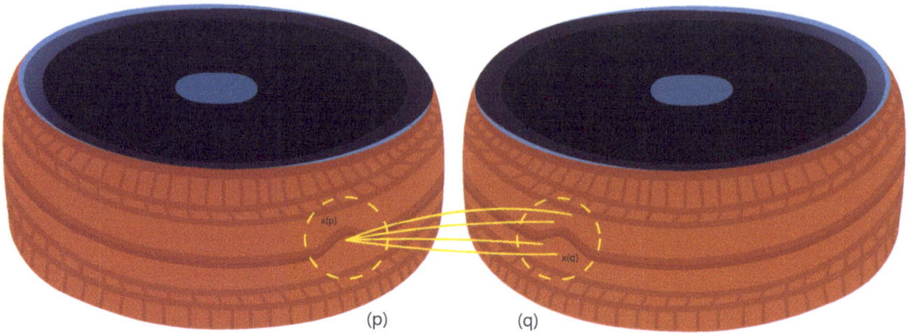

Fig. 1.2. Consider the immediate neighbourhoods of the deformation points p and q in tyres A and B. The deformation at point p in A has a *one-to-one* connection to another deformation point q in B and it has an *onto* connection to multiple others in the neighbourhood of q, i.e., in $x(q)$. The continuous and invertible function f that maps points in A to B, i.e., $f : A \to B$, is called a homeomorphism.

ized points p and q are *non-interfering*. Also, in the collection of the deformations of all tyres, there are many which might belong to tyre B. Mathematically, this is called an *open set*. In topology, the car is defined to be the entire set C and the tyres A, B, C, and D are its subsets, C is an open set if it is in a subset. Hence, we can distinguish the sets of deformations around the points p and q by two non-interfering open sets, this property is called *Hausdorff*. Moreover, each tyre comes with a family of deformations, each which are unique and universally belong to the entire set C and are homeomorphic to other deformations in different tyres and hence to the whole set of deformations. Thus, these deformation families constitute the entire set C, i.e., $C = \cup_\alpha \nu_\alpha$, where ν_α is a subset (maybe A, B, C or D depending upon α) and the homeomorphisms are depicted using maps, $\phi_\alpha : \nu_\alpha \to R^n$, where R^n is the set of all Euclidean closed geometries in a Euclidean space of dimension n. If the above conditions are satisfied, i.e. if a topological space is Hausdorff, and comes with a family $\{(\nu_\alpha, \Phi_\alpha)\}$ with set ν_α being a subset of an open set C and homeomorphisms $\Phi_\alpha : \nu_\alpha \to R^n$ such that $C = \cup_\alpha \nu_\alpha$, we call such a topological space C to be an n-dimensional smooth manifold. The pairs $(\nu_\alpha, \Phi_\alpha)$ are called *charts*, the family $\{(\nu_\alpha, \Phi_\alpha)\}$ is called an *atlas*, and Φ_α is called a *coordinate function.*

In topology, *smooth* corresponds to *differentiable*, and if our manifold has the metric signature, $(- \ + \ + +)$, we refer to it as a *Lorentzian manifold* (C, g), where C is our topological manifold and **g** is the metric tensor. A Lorentzian manifold is a type of *pseudo-Riemannian manifold* (C, g) which is a differentiable manifold C equipped with a non-degenerate, smooth, and symmetric metric tensor **g**. Spacetime is mathematically defined as a 4-dimensional, smooth, connected Lorentzian manifold (C, g). Hence, if our car is a smooth, 4D Lorentzian manifold then each localized deformation's frame of reference on this manifold is represented using coordinate charts. We adopt a more formal and mathematically rigorous, but intuitive approach to differential topology in the next chapter.

2

Manifolds: The One With The Rigorous Approach

Gravitational physics done in spacetime focuses on topological spaces (M, Ψ) that can be charted similar to how the surface of the Earth can be charted on an atlas. Thus, a topological space (M, Ψ) is known as an n-dimensional topological manifold if for all points d which belongs to the manifold M there not only exists an open set V in the topology Ψ, which contains the point d, but also exists an entire map z that takes every point in the set V to the subset in R^n in an invertible, one-to-one manner which in both directions is continuous. Representing this mathematically we write $\forall d \in M : \exists V \in q : \exists z : V \to z(V) \subseteq R^n$ Where $z(V)$ is the image of the domain under the chart z. Note that this mapping of the point is done in such a way that:
1. z is invertible, i.e., there exists a map z^{-1} such that: $z^{-1} : z(V) \to V$,
2. z is continuous, and
3. z^{-1} is continuous.

2.1 Chart

Going back to the car analogy, we observe that the tyre of the car is actually a torus (the shape of a doughnut). We can claim that this tyre surface is the set M which is equipped with some topology Ψ in R^3, i.e., $M \subseteq R^3$ (M is a subset of R^3 equipped with a topology Ψ). Now, since the tyre is elastic, it deforms at points when it encounters rocks on the terrain. For a deformation point p, there exists an open subset V (which is its immediate neighbourhood and whose boundary is decided based on to what extent the neighbourhood is affected due to the rock encounter), and there exists a map z which maps the deformation point p and every point in its neighbourhood (i.e., in its open subset V), to some part of R^2. This mapping done is bijective and generates an open region in the mapped part of R^2, which is nothing but a set of real numbers. Mathematically, the previous statement is expressed as follows: $R^2 = R \times R = \{(x, y)|x, y \in R\}$. Say the deformation p occurred due to a rock R_1, then we can say that the position of the deformation made by the rock R_1 in $M \subseteq R^3$ is at $z(p) \subseteq R^2$, which contains two components: $z(p) = (z^1(p), z^2(p))$. What this means is that the position of the deformation of the rock R_1 on the three-dimensional surface of the tyre is mapped to a two-dimensional region and identified using two coordinates $z^1(p)$ and $z^2(p)$. The pair (V, z) is called a chart, where V is the boundary of the deformation effect of the rock in R^3 and $z(V)$ is its image existing in R^2. The whole tyre is covered by charts that contain deformation points. Thus, for every point on the manifold, there exists a chart that contains the point, and the topological space (M, Ψ) is called a two-dimensional topological manifold.

2.2 Atlas

(V, z) is a chart of $(M, \Psi.)$ This manifold that contains a collection of charts that can be classified under a set A which mathematically is: $A = \{(V_\rho, z_\rho)|\rho \in I\}$ (where ρ is a label which belongs to some arbitrary index set I), is called an atlas of (M, Ψ). In other words, the atlas comprises of a family of charts. The existence of the atlas A is subject to the condition that the union of all the charts domains must reproduce the original surface M, i.e., $M = \cup_{\alpha \in I} V_\alpha$. What this means is that if we take all the images $z(V_\alpha)$ of all the open sets V_α existing in R^2 and stitch them together, we must be able to reproduce the surface of the tyre again. The topological space M is said to be *paracompact* if for every atlas $A = \{(V_\rho, z_\rho)|\rho \in I\}$ there exists a locally finite atlas $B = \{(V_\sigma, z_\sigma)|\sigma \in I\}$ with each open set U_σ contained in some V_ρ.

2.3 Chart and Coordinate Maps

In general, for an n-dimensional topological manifold the function $z : V \to z(V) \subseteq R^n$ is called a chart map. It is important to note that $R^n = R \times R \times ...$ is the set of n-tuples, thus, the image of a deformation point in R^n is represented using the coordinates $z(p) = (z^1(p), z^2(p), , z^n(p))$, where z^j is a map which takes a point in V and maps it to R (a real number), i.e., $z^j : V \to R$ for $j = 1, 2, , n$. A mathematical picture of this is as follows

$$z : V \to R^n = \left\{ \begin{array}{l} z^1 : V \to R \\ ... \\ z^n : V \to R \end{array} \right\} = z : V \to R^n = \textstyle\sum_{j=1}^{n} z^j,$$

where the individual charts z^j are called coordinate maps. What this means is that the deformation point $p(\in V)$ has its first coordinate at $z^1(p)$ present in the region $z(V)$ of the chart (V, z), its second coordinate at $z^2(p)$ present in the region $z(V)$ of the chart (V, z), and so on.

2.4 Chart Transition Maps

Consider two charts (V, z) and (U, w), with overlapping regions on the surface M, equipped with a topology $\varPsi$. Let the tyre encounter an arbitrary but small distribution of rocks (of the same shape and mass), the deformation points are all alike and exist within a region on the tyre. Let's call this the *deformation region.* Consider the points to the *deformation region's* immediate right and immediate left. When the tyre encounters the rock distribution, the regions to the left and right are affected to some extent (whose boundary is set based on the magnitude of deformation). Let the open set of the left region containing the set of affected points be V and let the open set of the right region containing the set of affected points be U. Thus, we can conclude that the open sets contain a non-empty overlap (which is the *deformation region* itself), i.e., $V \cap U \neq \emptyset$. We know that V comes with a chart map z that takes any point in V and maps it to some region in R^n, i.e., $z : V \to z(V) \subseteq R^n$, and U contains with a chart map w that takes any point in U and maps it to some other region in R^n, i.e., $w : U \to w(U) \subseteq R^n$. Now a deformation point d present in the *deformation region* (which is the intersection of V and U) can be mapped to two regions of R^n using the chart maps z and w. Similarly, we can map all the points in the *deformation region* $(V \cap U)$ into $z(V)$ and $w(U)$ via chart maps z and w to obtain regions $z(V \cap U)$ and $w(V \cap U)$ in R^n. What this implies is that a point d in the *deformation region* $(V \cap U)$ is mapped to two points in the regions $z(V \cap U)$ and $w(V \cap U)$ in R^n. A natural question that arises is- how are these two points related? Let the mapped point of d in $z(V \cap U)$ be d', since we know that z is invertible we use this property to define d'', the mapped point of d in $w(V \cap U)$ as follows: $d'' = w(z^{-1}(d')) = w \circ z^{-1}(d')$. Thus, $w \circ z^{-1}$ acts as a chart map which maps points of $z(V \cap U)$ existing in R^n to $w(V \cap U)$ existing in R^n. Formally, this is known as a chart transition map. Chart transition maps contain the information of how to stitch together all the charts of an atlas.

Consider the tyre again of surface M equipped with a topology Ψ, now we specify that it is the Euclidean space of n-dimensions, R^n. R^n is nothing but the set of all n-tuples $(x^1, x^2, ..., x^n)$, with $-\infty < x^j < \infty$. Let $\frac{R^n}{2}$ be the 'lower half' of R^n, i.e., the lower half of the tyre for which $x^1 \leq 0$. Let p be the point existing on the lower half of the tyre and let V be the open set in which it is contained. The map z of the open set $V \subset R^n$ (in $\frac{R^n}{2}$) to the open set $V' \subset R^m$ (in $\frac{R^m}{2}$) is said to be of class C^k if the coordinates of the image point $z(p) = (z^1(p), z^2(p), ..., z^m(p)) = (x'^1, x'^2, ..., x'^m)$ in V' are k-times continuously differentiable functions (which refers to the existence of the kth derivative which is continuous) of the coordinates $(x^1, x^2, ..., x^n)$ of the point p in the open set V on the tyre. A map is called C^∞ if it is C^k for all $k \geq 0$ and C^0 if it is continuous.

A C^k n-dimensional manifold M is a set M together with a C^k atlas $A = \{(V_\rho, z_\rho)|\rho I\}$ where V_ρ are subsets of M and $z_\rho : V_\rho \rightarrow z_\rho(V_\rho) \subseteq R^n$ are the one-to-one maps such that all the open sets V_ρ cover M. If there exists a non-empty overlap between two open sets V_ρ (with map z_ρ) and U_σ (with map w_σ), i.e., $V_\rho \cap U_\sigma \neq \emptyset$, then the map $z_\rho \circ w_\sigma^{-1} : w_\sigma(V_\sigma \cap U_\sigma) \rightarrow z_\rho(V_\rho \cap U_\sigma)$ is a C^k map of an open subset of R^n to an open subset of R^n.

2.5 Homeomorphism and Diffeomorphism

Let the images of V and U existing in R^n be v and u, i.e., let $z(V) = v$, and $w(U) = u$, and stated previously these are open sets since the chart maps are invertible. A mapping between the open sets of R^n, $f : v \to u$, is called a homeomorphism if it is bijective and if f and its inverse f^{-1} are continuous. A differential homeomorphism is called a diffeomorphism. Analogous to how a homeomorphism is a bijection that is continuous and also possesses a continuous inverse, a diffeomorphism is a bijection which is differentiable with a differentiable inverse, i.e., if v and u are connected open subsets of R^n such that u is simply connected, a differentiable map $f : v \to u$ is a diffeomorphism if the differential $Df_p : R^n \to R^n$ is bijective at each point p in v. Another way to put this is to state that a mapping f between open sets of $R^n : v \to u$ is called a diffeomorphism if it is bijective and if f and its inverse mapping f^{-1} are differentiable. Hence, every diffeomorphism is a homeomorphism, but not vice-versa. Generally, a bijective mapping is a C^k diffeomorphism if f and f^{-1} are of class C^k (see table). Thus, the map z from M to M' is said to be a C^k diffeomorphism if it is a one-one C^k map and the inverse z^{-1} is a C^k map from M' to M. We observed that the set of all deformations around an arbitrary point p on tyre A was not similar to that around point q on the tyre B, in other words, the sets of deformations around localized points p and q are non-interfering. Also, in the collection of the deformations of all tyres, there are many which might belong to tyre B. Consider a curve η present on the manifold M. This curve η can be called k-times continuously differentiable if there exists a C^k atlas.

If you are comfortable with the C^k classes, then, another definition to diffeomorphisms can be adopted. Two manifolds M, N are said to be diffeomorphic if there exists a homeomorphism $f : M$ such that f is a C^∞ function with a C^∞ inverse. f is called a diffeomorphism.

Mathematically, we know that this is called an open set. In topology, the car is defined to be the entire set C and the tyres $A, B, C,$ and D are its subsets, C is an open set if it is in a subset. Hence, we can distinguish the sets of deformations around the points p and q by two non-interfering open sets, this property is called *Hausdorff*. Thus, in short, a topological space M is said to be a Hausdorff space if for two points a and b in M, there exists disjoint open sets V and U in M such that $a \in V$ and $b \in V$. This condition is sometimes called the *Hausdorff separation axiom* .

Atlas	Properties
C^0	$C^0\,(R^n \to R^n)$ are continuous maps
C^1	$C^1\,(R^n \to R^n)$ are the maps that are once differentiable and continuous
C^k	$C^k\,(R^n \to R^n)$ are the maps that are k-times continuously differentiable
D^k	$D^k\,(R^n \to R^n)$ are the maps that are k-times differentiable but are not continuous
C^∞	$C^\infty\,(R^n \to R^n)$ are the maps that are many-times continuously differentiable
$C^{'\infty}$	$C^{'\infty}\,(R^n \to R^n)$ are the maps that are many-times continuously complex differentiable
C^ω	C^ω are the maps that can be Taylor expanded

Table 2.1. This table depicts the properties of class k atlases. $C^{'\infty}$ this is valid only for even dimensional manifolds under the condition that the chart maps satisfy the Cauchy-Riemann equations . C^ω stands for analytic; a function $f : R^n \to R$ is analytic at $p \in R^n$ if f can be expressed as a power series in the $(x^j - p^j)$ which converges in some neighbourhood of p.

Moreover, each tyre comes with a family of deformations, each which are unique and universally belong to the entire set M and are homeomorphic to other deformations in different tyres and hence to the whole set of deformations. Thus, these deformation families constitute the entire set M, i.e., $M = \cup_{\rho \in I} V_\rho$, where V_ρ is a subset (maybe A, B, C or D) and the homeomorphisms are depicted using maps, $z_\rho : V_\rho \to zV_\rho \subseteq R^n$, where R^n is the set of all Euclidean closed geometries in a Euclidean space of dimension n. If the above conditions are satisfied, i.e. if a topological space is Hausdorff, and comes with a family $\{(V_\rho, z_\rho) | \rho \in I\}$ with set V_ρ being a subset of an open set M and homeomorphisms $z_\rho : V_\rho \to z(V_\rho) \subseteq R^n$ such that $M = \cup_{\rho \in I} V_\rho$, we call such a topological space M to be an n-dimensional smooth manifold. All manifolds considered are assumed to paracompact, connected C^∞ Hausdorff manifolds without boundary.

2.6 Differential Manifold

Knowing the concept of a diffeomorphism, we can now reframe the concept of a topological manifold. An atlas bequeaths M with the structure of a topological manifold, of dimension n, if the mappings $w_\sigma \circ z_\rho^{-1}$ are homeomorphisms (i.e., continuous bijections) between open sets of R^n, namely between $z_\rho(V_\rho \cap U_\sigma)$ and w_σ $(V_\rho \cap U_\sigma)$. If these mappings are diffeomorphisms, then the manifold can be called a differential manifold. Generally, the manifold is a differential manifold of class C^k if these mappings are C^k diffeomorphisms. Thus, the term smooth means that a class C^k with k large enough (in particular $k = \infty$). A differential manifold (or a smooth manifold) is often written as a C^∞ manifold. Thus, we can define a C^∞ manifold as to be a pair (M, A), where A is a maximal atlas for M.

Consider two C^∞ manifolds, (R, V) and (R, U). These manifolds are called isomorphic if there exists a bijective (one-to-one and onto) function $f : R \to R$ such that $p \in V$ if and only if $p \circ f \in U$. Two C^∞ manifolds (M, A) and (M', B) are called *diffeomorphic* if there is a bijective function $f : M \to M'$ such that $p \in B$ if and only if $p \circ f \in A$. Few books refer to the atlas for M as the *differential structure* for M.

2.7 Tangent Space and Tangent Bundle

We take the tyre again and connect all the points across the threads which appear to possesses information regarding the deformations caused to the tyre. We connect the points (or events) with a continuous and smooth curve P. Upon examination, we observe that events on the curve occur at regular intervals (assumption made for simplicity) and using this fact we parameterize the curve in terms of χ. Hence the parameterized curve takes the form of $\frac{dP}{d\chi}$. Consider the very first event on the tyre Q and the very last event R. If the points were present within the same thread, then the curve $P(\chi)$ would actually be the straight line described by the equation-$P(\chi) = X + \chi(R - Q)$. The derivative of $P(\chi)$ can be formed as follows

$$\frac{d}{d\chi}(X + \chi(R - Q)) = \mathbf{R} - \mathbf{Q} = v_{QR} = \left(\frac{dP}{d\chi}\right)_{\chi=0} \tag{2.1}$$

This is defined to be a tangent vector. More formally, a tangent vector v to the differential manifold M at a point $p \in M$ is defined as $(V_\rho, z_\rho, v_{z_\rho})$, where (V_ρ, z_ρ) are charts which contain p and $v_{z_\rho} = v^j_{z_\rho}, j = 1, 2, ..., n$ are vectors in R^n. What is more interesting is the resting place of this vector. This tangent vector does not lie on the manifold, i.e., it does not share the same home as that of the cure $P(\chi)$, rather it lies in a so-called tangent space which touches or makes contact with the manifold only at $P(\chi = a)$, the point where $\frac{dP}{d\chi}$ was evaluated. Imagine that we take different colours of moulding clay and press them against the tyre, starting at specific points (events). Thus, all the tangent vectors of every event will be contained in specific bits of clay. Now, mould all these pieces of clay onto a single, larger, and continuous piece of clay. This is the tangent space which is a plane in which all the tangent vectors to all events are contained such that the plane is tangent to the tyre at every point.

All the tangent vectors at the point p constitute a tangent space (which is a vector space) to M^n at the point p. This tangent space is denoted by T_pM^n or simply a $T_p(M^n)$. A tangent bundle is defined as the set of the pairs of the points and the tangent vector of that point, i.e., (p, v_p), where $p \in M^n$ (point present in the n-dimensional manifold) and $v_p \in T_pM^n$ (tangent vector contained in the tangent space of the n-dimensional manifold), denoted by TM^n.

2.8 Immersions and Embeddings

An immersion is defined to be the function between differential manifolds whose derivative is everywhere injective (one-to-one), i.e., the function $h : M \to M'$ is called an immersion between M and M' (differential manifolds) if $D_ph : T_pM \to T_{f(p)}N$ is an injective function at every point p of M. In other words, an immersion simply means that the tangent spaces are mapped injectively, i.e., the map described above is injective.
Consider the map z from a C^k n-dimensional manifold M to a C^l o-dimensional manifold N. This map is called a C^r map $(r \leq k, r \leq l)$ if, for the coordinates of the image point $z(p)$ in N are C^r functions of the coordinates of p in M. A C^r map z $(r \geq 0)$ is called an immersion if it and its inverse are C^r maps, i.e., if for each point $p \in M$ there exists an open set V such that the inverse map z^{-1}

restricted to the image of the domain $z(V)$.

An immersion is called an embedding if it is a homeomorphism onto its image in the topology Ψ of a differentiable manifold M. It is important to note that all embeddings are one-to-one immersions, however, the converse is not true.

2.9 Pseudo-Riemannian metrics

A metric **g** on a manifold M which is a symmetric covariant 2-tensor field is called a pseudo-Riemannian metric if the determinant $|g|$ with elements $g_{\alpha\beta}$, whose quadratic form it defines on contravariant vectors, $g(A, B)$, given in local charts by $g_{\alpha\beta}A^{\alpha}B^{\beta}$, does not vanish in any chart, i.e., it is non-degenerate. This definition is independent of the choice of charts because under a change of local coordinates $(x'^{m}) \rightarrow (x^{m})$ it holds that

$$|g| = |g'|\frac{dx'}{dx} \tag{2.2}$$

(M, g) is a diffeomorphism f which leaves **g** invariant, i.e., $f^{*}g = g$. Two pseudo-Riemannian manifolds (M, g) and (M', g') are called locally isometric if there exists a differential mapping f such that any point $p \in M$ admits a neighbourhood M, and $f(p)$ a neighbourhood M' with (M, g) and (M', g') isometric. It is important to note that pseudo-Riemannian manifolds can have different topologies although they possess the same dimension. Flat space is defined as a pseudo-Riemannian manifold is isometric with a pseudo-Euclidean space.

2.10 Lorentzian Manifold

In topology, *smooth* corresponds to *differentiable*, and if our manifold has the metric signature, $(- \ + \ + \ +)$, we refer to it as a Lorentzian manifold (M, g), where M is our topological manifold and **g** is the metric tensor. A Lorentzian manifold is a type of pseudo-Riemannian manifold -Riemannian manifold (M, g) which is a differentiable manifold M equipped with a non-degenerate, smooth, and symmetric metric tensor **g**.

Spacetime is mathematically defined as a 4-dimensional, smooth, connected Lorentzian manifold (M, g). Here M is a connected 4-dimensional Hausdorff C^∞ manifold and **g** is a Lorentz metric on M. Hence, if our car is a smooth, 4-D Lorentzian manifold then each localized deformation's frame of reference on this manifold is represented using coordinate charts. Similarly, in the spacetime manifold, the coordinate charts are used to represent observers in reference frames. For a physicist, the most preferred and useful definition is by identifying (locally) the manifold by R^n.

2.11 Whitney and Nash Embedding Theorems

In short, the idea of the theorem is that any $C^{k\geq 1}$ atlas A of a topological manifold contains a C^∞ manifold, i.e., an atlas in which the chart transitional maps are at least once continuously differentiable we can *remove* more and more charts until we are left with a C^∞ atlas. What this implies is that we may always consider C^∞ manifolds (called *smooth manifolds*) from now on.
Let's formulate this mathematically. Let M be a smooth topological manifold of dimension n. The theorem roughly states that if M^n is a compact C^∞ manifold, then there is an embedding $z : M \to R^N$ for some N. The strongest version of the theorem is given below (without proof).

Theorem 2.1. *Any smooth manifold of dimension n can be immersed into R^{2n-1} and embedded into R^{2n}*

Nash embedding theorem states that every Riemannian manifold can be isometrically embedded into some Euclidean space. Why this is interesting is because it mentions isometric embedding, i.e. preserving the length of curves in the manifold, whereas the Whitney theorem does not. According to this theorem, 4-dimensional curved spacetime can be isometrically embedded in a flat spacetime of 39 dimensions or less[1].

[1] For the mathematically inclined see: Nash, J. (1954). C1 isometric imbeddings. Annals of mathematics, 383-396, and Nash, J. (1956). The imbedding problem for Riemannian manifolds. Annals of mathematics, 20-63.

2.12 Einsteinian Spacetime

A metric is called Riemannian if it's quadratic form, **g**, is positive definite. A pseudo-Riemannian metric **g** can be called a Lorentzian metric if the sign of the **g** is $(- \ + \ ... \ +)$. A spacetime of General Relativity is a pair (M, g), where M is a differentiable manifold and **g** is a Lorentzian metric on M. Such a spacetime is called Einsteinian if there exists a physically meaningful stress-energy tensor (of rank two) **T** such that the following equations are satisfied on the differentiable manifold M: $Einstein(g) = T$, and $\nabla T = 0$[2].

[2] After reading the last chapter on Einstein's field equations, you would realize that the first equation corresponds to $G^{\mu\nu} = \frac{8\pi G}{c^4} T^{\mu\nu}$, and the second equation talks of the conservation of energy which is just a consequence of Einstein's field equations

3

Differential Forms and Tensors

3.1 1-Forms

A 1-form $\boldsymbol{\alpha}$ is a linear, real-valued function of vectors. Consider a point p present in the manifold M equipped with a topology Ψ. If $\mathbf{R}$ is a vector at p, the number into which the 1-form maps $\mathbf{R}$ is expressed as $\langle \boldsymbol{\alpha}, \mathbf{R} \rangle$. This is nothing but the value of $\boldsymbol{\alpha}$ on $\mathbf{R}$ or simply, the contraction of $\boldsymbol{\alpha}$ on $\mathbf{R}$. For the sake of simplicity, let's call this the *carrot operator* and refer to its operation as *carroting*[1]. In other words, we can define a 1-form $\boldsymbol{\alpha}$ at a point p in (M, Ψ) as a linear, real-valued function on the tangent space T_p of the vectors at p. The condition of linearity allows us to arrive at two conclusions, the first one is to realize the mathematical property of linearity, and the other is to form a visual understanding via surfaces.

Linearity implies the following

$$\langle \boldsymbol{\alpha}, a\mathbf{R} + b\mathbf{S} \rangle = a \langle \boldsymbol{\alpha}, \mathbf{R} \rangle + b \langle \boldsymbol{\alpha}, \mathbf{S} \rangle \tag{3.1}$$

The tangent space T_p, for a given 1-form $\boldsymbol{\alpha}$, defined by $\langle \boldsymbol{\alpha}, \mathbf{R} \rangle = const$, is linear. We can imagine a 1-form as a set of planes in the tangent space. Imagine that each of these planes had a high-sensitive *vector-alarm* and as a vector pierces through a plane, the alarm would go off (which will be recorded). Also let's make the assumption that each alarm makes a unique sound (thus, enabling

[1] The credit for naming this operator goes to my students of the Gravitational Summer School '18

us to distinguish the panes). When $\langle \boldsymbol{\alpha}, \mathbf{R} \rangle = 0$, the vector (or more precisely, it's tip) touches the first plane and we hear the first alarm. Similarly when $\langle \boldsymbol{\alpha}, \mathbf{R} \rangle = n$, the vector pierces through the $(n+1)^{th}$ plane and in total we hear $(n+1)$ distinct alarm sounds.

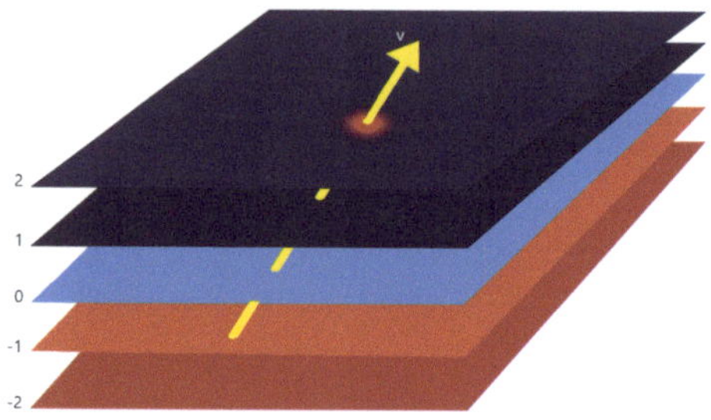

Fig. 3.1. Each of these planes have a high-sensitive *vector-alarm* and as a vector pierces through a plane, the alarm goes off. In this figure, the vector **v** pierces through the 1-form $\boldsymbol{\sigma}$ and $\langle \boldsymbol{\sigma}, \mathbf{v} \rangle = 4$, i.e., we hear 4 distinct alarms go off.

The simplest example of a 1-form is the gradient **dg** of a function **g**. Consider a vector u, and a curve $P(\eta)$ ($P(\eta) = \eta u + P_0$ which is parameterized in terms of η), and differentiate the function **g** along this curve.

$$\partial_{\mathbf{u}} g = \left(\frac{d}{d\eta} \right)_{\eta=0} g\left[P(\eta)\right] = \left(\frac{\mathbf{dg}}{d\eta} \right)_{P_0} \tag{3.2}$$

Observe that the operators $\partial_{\mathbf{u}} = \left(\frac{d}{d\eta} \right)_{\eta=0, P(\eta)}$, are related, i.e., the directional derivative and the gradient are related. Let the surfaces present in the tangent space T_p defined for the point p in (M, Ψ) be numbered with respect to **g**, (i.e., $g = 1$: surface one; $g = 2$: surface two; ...). Let the initial position of the vector (which starts from some arbitrary surface of **g**) be P_0. The first point of contact of the vector with a surface would be given as $\langle \mathbf{dg}, P - P_0 \rangle$, where **dg** is the stack of infinitesimal surfaces present between two **g**-surfaces. Thus, a generalized expression can be obtained: $g(P) = g(P_0) + \langle \mathbf{dg}, P - P_0 \rangle$. Since the relation between the directional derivative and the gradient is well established, let's apply $\partial_{\mathbf{u}}$

to $g(P)$ and evaluate the result at the take-off point P_0.

$$\partial_{\mathbf{u}} g = \left\langle \mathbf{dg}, \frac{dP}{d\eta} \right\rangle = \langle \mathbf{dg}, \mathbf{u} \rangle \tag{3.3}$$

In general $g(P)$ will have non-linear contributions of the order $O(P - P_0)$. Similar to how vectors possesses a basis $\mathbf{e}_\beta$, 1-forms possess basis too. These basis 1-forms are denoted by $\boldsymbol{\omega}^\alpha$.

Lemma 3.1. *If the basis 1-forms set defined by $\boldsymbol{\omega}^\alpha$ and the basis vectors defined by $\mathbf{e}_\beta$ are duals of each other, then $\langle \boldsymbol{\omega}^\alpha, \mathbf{e}_\beta \rangle = \delta^\alpha_\beta$*

This lemma enables us to expand any arbitrary vector and 1-form in terms of their basis as follows: $e_\alpha, u = u^\alpha \mathbf{e}_{\boldsymbol{\alpha}}$ and $\rho = \rho_\beta \boldsymbol{\omega}^{\boldsymbol{\beta}}$. We shall calculate the surfaces of $\boldsymbol{\rho}$ pierced by a basis vector $\mathbf{e}_\alpha$, and also calculate the carroting $\langle \boldsymbol{\omega}^{\boldsymbol{\alpha}}, \mathbf{u} \rangle$ for a vector $\mathbf{u} = \mathbf{e}_{\boldsymbol{\beta}} u^\beta$.

The piercing example
$\langle \boldsymbol{\rho}, \mathbf{e}_{\boldsymbol{\alpha}} \rangle = \left\langle \boldsymbol{\rho}_{\boldsymbol{\beta}} \boldsymbol{\omega}^{\boldsymbol{\beta}}, \mathbf{e}_{\boldsymbol{\alpha}} \right\rangle = \rho_\beta \left\langle \boldsymbol{\omega}^{\boldsymbol{\beta}}, \mathbf{e}_{\boldsymbol{\alpha}} \right\rangle = \rho_\beta \delta^\beta_\alpha = \rho_\alpha$

The carroting example
$\langle \boldsymbol{\omega}^{\boldsymbol{\alpha}}, \mathbf{u} \rangle = \left\langle \boldsymbol{\omega}^{\boldsymbol{\alpha}}, \mathbf{e}_{\boldsymbol{\beta}} u^\beta \right\rangle = u^\beta \left\langle \boldsymbol{\omega}^{\boldsymbol{\alpha}}, \mathbf{e}_{\boldsymbol{\beta}} \right\rangle = u^\alpha$

Well what's the bigger picture here? Go on and multiply the piercing example with u^α, the carroting example with ρ_α and add both the equations to obtain the following result.

$$\begin{aligned}
&[\langle \boldsymbol{\rho}, \mathbf{e}_{\boldsymbol{\alpha}} \rangle u^\alpha + \langle \boldsymbol{\omega}^{\boldsymbol{\alpha}}, \mathbf{u} \rangle \rho_\alpha] \\
&[\langle \boldsymbol{\rho}, \mathbf{e}_{\boldsymbol{\alpha}} u^\alpha \rangle + \langle \rho_\alpha \boldsymbol{\omega}^{\boldsymbol{\alpha}}, \mathbf{u} \rangle] = \rho_\alpha u^\alpha + \rho_\alpha u^\alpha \\
&[\langle \boldsymbol{\rho}, \mathbf{u} \rangle + \langle \boldsymbol{\rho}, \mathbf{u} \rangle] = 2\rho_\alpha u^\alpha \\
&\langle \boldsymbol{\rho}, \mathbf{u} \rangle = \rho_\alpha u^\alpha
\end{aligned} \tag{3.4}$$

Thus, we have obtained a way of using components to calculate the coordinate-independent value of $\langle \boldsymbol{\rho}, \mathbf{u} \rangle$. Let's discuss a bit more on what the *dual* is (from the lemma). Since we can express the 1-form $\boldsymbol{\alpha}$ at a point p in terms of its basis $\boldsymbol{\alpha} = \boldsymbol{\omega}^{\mathbf{j}} \boldsymbol{\alpha}_{\mathbf{j}}$, the set of all 1-forms at p forms an n-dimensional vector space at p. This vector space is called the *dual space* of the tangent space T_p and is written as ${}^*T_p$. Let's revise the lemma a bit and re-state it as follows

Lemma 3.2. *For any 1-form* $\boldsymbol{\alpha} \in {}^*T_p$ *and any vector* $\mathbf{R} \in T_p$*, we can express the carroting* $\langle \boldsymbol{\alpha}, \mathbf{R} \rangle$ *in terms of the corresponding dual basis* $\boldsymbol{\omega}^{\mathbf{j}}$ *and* $\mathbf{e_j}$ *by relations*

$$\langle \boldsymbol{\alpha}, \mathbf{R} \rangle = \left\langle \boldsymbol{\alpha}_{\mathbf{j}} \boldsymbol{\omega}^{\mathbf{j}}, \mathbf{R}^{\mathbf{j}} \mathbf{e_j} \right\rangle = \alpha_j R^j$$

Each function $\mathbf{g}$ on the manifold M defines a 1-form at a point p. This follows a rule which states that for each vector R, $\langle \mathbf{dg}, \mathbf{R} \rangle = \mathbf{Rg}$. Here, dg is called the differential of $\mathbf{g}$.

3.2 Two Roads to Tensors: Road for Pedestrians

A tensor is like a slot machine, not just any ordinary one but rather a very modified machine. The generalized tensor slot machine has two slots (instead of one in the real machine) and two sub-slots. These slots are specific in what they accept. There are n first sub-slots and it accepts only 1-forms while there are m second sub-slots which accept only vectors. Thus, we can mathematically represent a tensor slot machine S as follows

$$S(\underbrace{\boldsymbol{\alpha}, \boldsymbol{\beta}, \boldsymbol{\gamma}, ..., \boldsymbol{\zeta}}_{n\ 1-forms}, \underbrace{\mathbf{u}, \mathbf{v}, ..., \mathbf{a}}_{m\ vectors}) \tag{3.5}$$

This S tensor is said to be of rank $\begin{pmatrix} \mathrm{n} \\ \mathrm{m} \end{pmatrix}$. It is important to note that most of the tensors do not remain the same if two slots of either 1-forms or vectors or both are interchanged, i.e., $S(\boldsymbol{\alpha}, \boldsymbol{\beta}, \mathbf{v}, \mathbf{u}) \neq S(\boldsymbol{\beta}, \boldsymbol{\alpha}, \mathbf{u}, \mathbf{v})$. Consider the following example which demonstrates how to work with tensors.

Working with tensors Let F be a tensor of rank $\begin{pmatrix} 3 \\ 2 \end{pmatrix}$. Define the tensor by inserting the basis vectors of the 1-forms and the vectors as follows

$$\mathbf{F}^{\alpha\beta\gamma}_{\delta\eta} \equiv F\left(\boldsymbol{\omega}^{\alpha}, \boldsymbol{\omega}^{\beta}, \boldsymbol{\omega}^{\gamma}, \mathbf{e}_{\delta}, \mathbf{e}_{\eta}\right) \tag{3.6}$$

Now, the output can be calculated for the given input as follows

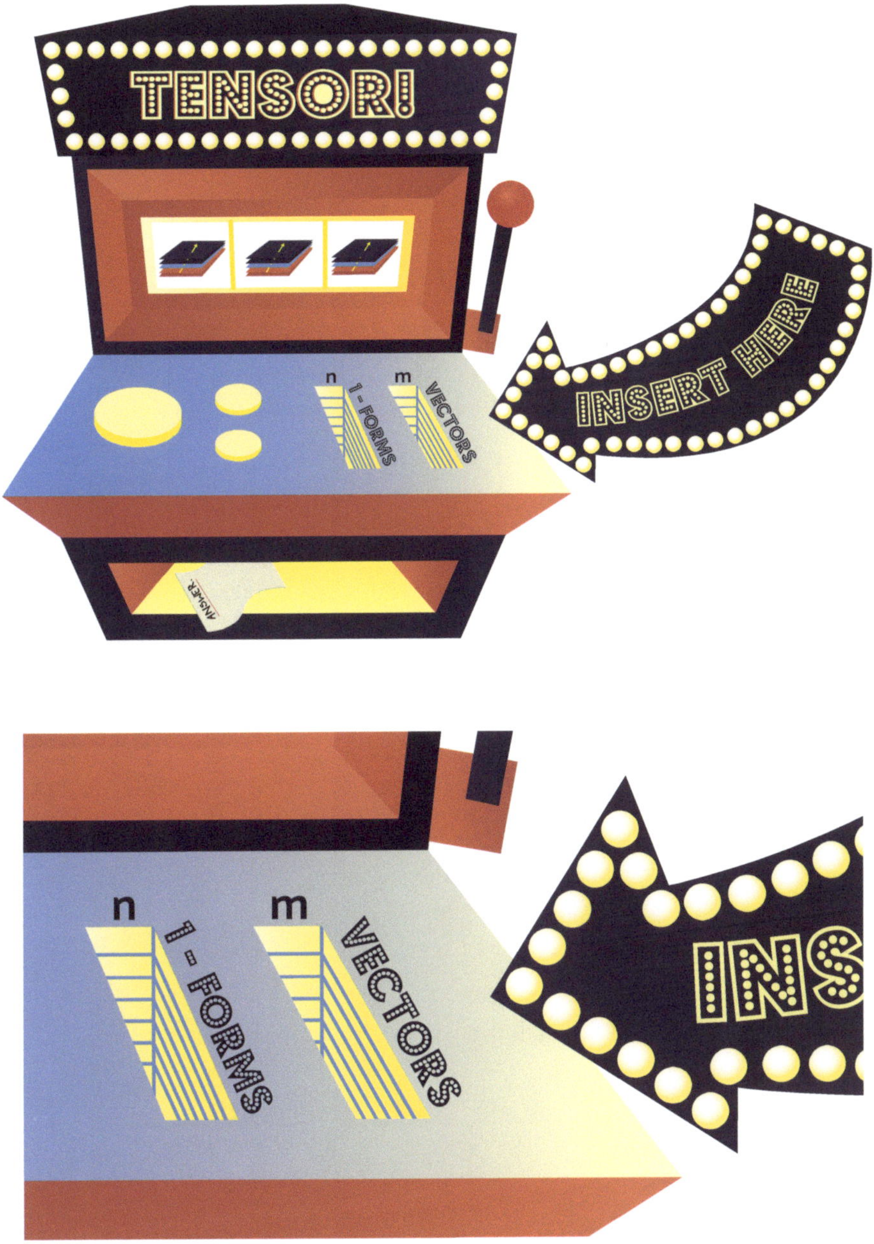

Fig. 3.2. The slot machine representation of tensors. The machine has n sub-slots accepting 1-forms and m sub-slots accepting vectors.

$$F\left(\boldsymbol{\sigma},\boldsymbol{\rho},\boldsymbol{\nu},\mathbf{u},\mathbf{v}\right)=F\left(\sigma_\alpha\boldsymbol{\omega}^\alpha,\rho_\beta\boldsymbol{\omega}^\gamma,\nu_\gamma\boldsymbol{\omega}^\gamma,u^\delta\mathbf{e}_\delta,v^\eta\mathbf{e}_\eta\right)$$
$$\sigma_\alpha\rho_\beta\nu_\gamma u^\delta v^\eta F\left(\boldsymbol{\omega}^\alpha,\boldsymbol{\omega}^\beta,\boldsymbol{\omega}^\gamma,\mathbf{e}_\delta,\mathbf{e}_\eta\right)=\mathbf{F}^{\alpha\beta\gamma}_{\delta\eta}\sigma_\alpha\rho_\beta\nu_\gamma u^\delta v^\eta \tag{3.7}$$

3.3 Two Roads to Tensors: Road for the Mathematically Inclined

We can form a *Cartesian product* from the tangent space T_p of vectors at point p and from the tangent space's dual $^*T_p$ of 1-forms at p as follows (The dual space $^*T_p$ is often called the *cotangent space.*)

$$\Pi^n_m=\underbrace{{}^*T_p\times{}^*T_p\times...\times{}^*T_p}_{n\ factors}\times\underbrace{T_p\times T_p\times...\times T_p}_{m\ factors} \tag{3.8}$$

this is the ordered set of 1-forms and vectors $(\boldsymbol{\alpha},\boldsymbol{\beta},\boldsymbol{\gamma},...,\boldsymbol{\zeta},\mathbf{u},\mathbf{v},...,\mathbf{a})$. A tensor of rank $\begin{pmatrix}\mathrm{n}\\\mathrm{m}\end{pmatrix}$ at a point p is a function on Π^n_m which is linear in each argument, i.e., if $\mathbf{T}$ is a tensor of rank $\begin{pmatrix}\mathrm{n}\\\mathrm{m}\end{pmatrix}$ at p, the number into which $\mathbf{T}$ maps the element $(\boldsymbol{\alpha},\boldsymbol{\beta},\boldsymbol{\gamma},...,\boldsymbol{\zeta},\mathbf{u},\mathbf{v},...,\mathbf{a})$ of Π^n_m as $T\left(\boldsymbol{\alpha},\boldsymbol{\beta},\boldsymbol{\gamma},...,\boldsymbol{\zeta},\mathbf{u},\mathbf{v},...,\mathbf{a}\right)$. The property of linerity also applies to tensors (this is described below).

$$T\left(\boldsymbol{\alpha},\boldsymbol{\beta},\boldsymbol{\gamma},...,\boldsymbol{\zeta},a\mathbf{R}+b\mathbf{S},\mathbf{u},\mathbf{v},...,\mathbf{a}\right)$$
$$=aT\left(\boldsymbol{\alpha},\boldsymbol{\beta},\boldsymbol{\gamma},...,\boldsymbol{\zeta},\mathbf{R},\mathbf{u},\mathbf{v},...,\mathbf{a}\right)+bT\left(\boldsymbol{\alpha},\boldsymbol{\beta},\boldsymbol{\gamma},...,\boldsymbol{\zeta},\mathbf{S},\mathbf{u},\mathbf{v},...,\mathbf{a}\right) \tag{3.9}$$

The space of all such tensors is called the tensor product .

$$\mathbf{T}^{\mathbf{n}}_{\mathbf{m}}(p)=\underbrace{{}^*T_p\otimes{}^*T_p\otimes...\otimes{}^*T_p}_{n\ factors}\otimes\underbrace{T_p\otimes T_p\otimes...\otimes T_p}_{m\ factors} \tag{3.10}$$

3.4 Tensor Operations: Addition

Let **S** and $\bar{S}$ be tensors of rank $\begin{pmatrix} \text{n} \\ \text{m} \end{pmatrix}$, the addition of these tensors is defined by the following rule

$$\begin{aligned} &\left(S+\bar{S}\right)(\boldsymbol{\alpha},\boldsymbol{\beta},\boldsymbol{\gamma},...,\boldsymbol{\zeta},\mathbf{u},\mathbf{v},...,\mathbf{a}) \\ &= S\left(\boldsymbol{\alpha},\boldsymbol{\beta},\boldsymbol{\gamma},...,\boldsymbol{\zeta},\mathbf{u},\mathbf{v},...,\mathbf{a}\right)+\bar{S}\left(\boldsymbol{\alpha},\boldsymbol{\beta},\boldsymbol{\gamma},...,\boldsymbol{\zeta},\mathbf{u},\mathbf{v},...,\mathbf{a}\right) \end{aligned} \tag{3.11}$$

3.5 Tensor Operations: Multiplication

The multiplication of the same tensor **S** considered in the previous example with a scalar ζ is shown below

$$\begin{aligned} &(\xi S)\left(\boldsymbol{\alpha},\boldsymbol{\beta},\boldsymbol{\gamma},...,\boldsymbol{\zeta},\mathbf{u},\mathbf{v},...,\mathbf{a}\right) \\ &= \xi \times S\left(\boldsymbol{\alpha},\boldsymbol{\beta},\boldsymbol{\gamma},...,\boldsymbol{\zeta},\mathbf{u},\mathbf{v},...,\mathbf{a}\right) \end{aligned} \tag{3.12}$$

A *covariant* k-tensor at a point $p \in M$ is defined as a k-multilinear form on k direct products of the tangent space T_pM. Similarly, a *contravariant* k-tensor at a point $p \in M$ is defined as a k-multilinear form on k direct products of the cotangent space $^*T_pM$.

The tensor product $\mathbf{S}\otimes\bar{\mathbf{S}}$ of an r-tensor **S** and as s-tensor $\bar{S}$ is an $(r+s)$-tensor with components defined by products of components. Consider the product of a covariant 2-tensor **T** and a contravariant 3-tensor $\bar{T}$. The result is a mixed 4-tensor $\mathbf{T}\otimes\bar{\mathbf{T}}$ with the following components

$$\left(\mathbf{T}\otimes\bar{\mathbf{T}}\right)_{\alpha\beta}^{\ \ \gamma\delta} = T_{\alpha\beta}\bar{T}^{\gamma\delta} = W_{\alpha\beta}^{\gamma\delta} \tag{3.13}$$

When we refer **T** as an r-covariant tensor and $\bar{\mathbf{T}}$ as an s-contravariant tensor it implies that they are elements of the tensor product of r copies of T_p and s copies of $^*T_p$. The covariant and contravariant tensors can also be defined in terms of local coordinates as follows

$$W_{\alpha\beta}^{\gamma\delta} = \underbrace{\left(\frac{\partial y^{\zeta}}{\partial x^{\alpha}}\frac{\partial y^{\eta}}{\partial x^{\beta}}\right)}_{covariant\ part} \underbrace{\left(\frac{\partial x^{\gamma}}{\partial y^{\rho}}\frac{\partial x^{\delta}}{\partial y^{\xi}}\right)}_{contravariant\ part} W_{\zeta\eta}^{\rho\xi} \tag{3.14}$$

Let's generalize this and write down the local coordinate transformation of an n-tensor.

$$S^{\alpha_1 \,\dots\, \alpha_m}_{\beta_1 \,\dots\, \beta_n}(x) = \frac{\partial y^{\mu_1}}{\partial x^{\beta_1}} \dots \frac{\partial y^{\mu_n}}{\partial x^{\beta_n}} \frac{\partial x^{\alpha_1}}{\partial y^{\nu_1}} \dots \frac{\partial x^{\alpha_m}}{\partial y^{\nu_m}} S^{\nu_1 \,\dots\, \nu_m}_{\mu_1 \,\dots\, \mu_n}(y) \quad (3.15)$$

3.6 Tensor Operations: Contraction

Recollect the slot-machine definition of a tensor. Contraction is similar to shutting off of sub-slots (which contain both 1-forms and vectors) within the two main slots. Consider the mixed 4-tensor $\mathbf{R} = R(\boldsymbol{\sigma}, \mathbf{u}, \mathbf{v}, \mathbf{w})$ which is of rank $\binom{1}{3}$. We can shut off sub-slots 1 and 3 to reduce the tensor to a covariant 2-tensor, say $\mathbf{S}$. This operation is described below.

$$S(u, w) = \textstyle\sum_{\alpha=0}^{3} R(\boldsymbol{\omega}^{\boldsymbol{\alpha}}, u, \mathbf{e}_{\boldsymbol{\alpha}}, w)$$

$$S(u, v) = S_{\mu\nu} u^\mu w^\nu = R^{\alpha}_{\mu\alpha\nu} u^\mu w^\nu \quad (3.16)$$

$$S_{\mu\nu} = R^{\alpha}_{\mu\alpha\nu}$$

3.7 Tensor Operations: Symmetrization and Antisymmetrization

If the output of a tensor is unaffected by an interchange of 2 input vectors or 1-forms, then it is called a symmetric tensor, if not then it is called an antisymmetric tensor.

Symmetric

$$T(\boldsymbol{\alpha}, \boldsymbol{\beta}, \boldsymbol{\gamma}) = T(\boldsymbol{\beta}, \boldsymbol{\alpha}, \boldsymbol{\gamma}) = T(\boldsymbol{\gamma}, \boldsymbol{\beta}, \boldsymbol{\alpha}) = \dots$$

(3.17)

Antisymmetric

$$T(\boldsymbol{\alpha}, \boldsymbol{\beta}, \boldsymbol{\gamma}) = -T(\boldsymbol{\beta}, \boldsymbol{\alpha}, \boldsymbol{\gamma}) = +T(\boldsymbol{\gamma}, \boldsymbol{\beta}, \boldsymbol{\alpha}) = - + \dots$$

Let $\mathbf{F}$ be a 2-covariant tensor. The symmetrization and antisymmetrization of $\mathbf{F}$ can be represented as follows

$$\begin{aligned} &Symmetrization: \; F_{(\alpha\beta)} = \tfrac{1}{2}\left(F_{\alpha\beta} + F_{\beta\alpha}\right) = S_{\alpha\beta} \\ &Antisymmetrization: \; F_{[\alpha\beta]} = \tfrac{1}{2}\left(F_{\alpha\beta} - F_{\beta\alpha}\right) = A_{\alpha\beta} \end{aligned} \tag{3.18}$$

3.8 Tensor Operations: Wedge Product

Given any two vectors, we can construct their *bivector* by *wedging* them. This can also be done with 1-forms to obtain 2-*forms*. This concept can also be used to construct a *trivector* and 3-*forms*.

$$\begin{aligned} &Bivector \\ &\mathbf{a} \wedge \mathbf{b} \equiv \mathbf{a} \otimes \mathbf{b} - \mathbf{b} \otimes \mathbf{a} \\ &2 - form \\ &\boldsymbol{\alpha} \wedge \boldsymbol{\beta} \equiv \boldsymbol{\alpha} \otimes \boldsymbol{\beta} - \boldsymbol{\beta} \otimes \boldsymbol{\alpha} \end{aligned} \tag{3.19}$$

This operator is interesting geometrically as it serves as a test of coplanarity[2]. Observe that the action of the wedge can be generalized- a wedge among p-forms produce a $(p+1)$-form. This raises a question- what maps these p-form fields to $(p+1)$-form fields, it this mapping linear?

3.9 Exterior differentiation

[3] The job of linearly mapping p-form fields to $(p+1)$-form fields is done by the exterior derivative d. If $z : M \to N$ is a C^r map

[2] Consider 3 arbitrary vectors $\mathbf{a}, \mathbf{b}, \mathbf{c}$. If $\mathbf{a}$ and $\mathbf{b}$ are collinear, then $\mathbf{a} = \mathbf{b}\lambda$. This implies that $\mathbf{a} \wedge \mathbf{b} = \mathbf{b}\lambda \wedge \mathbf{b} = \mathbf{b}\lambda \otimes \mathbf{b} - \mathbf{b} \otimes \mathbf{b}\lambda = 0$. Now, if $\mathbf{c}$ is coplanar with $\mathbf{a}$ and $\mathbf{b}$, then $\mathbf{c}$ can be expressed as a linear combination (i.e., scalar multiplication followed by vector addition) of the other two vectors, i.e., $\mathbf{w} = \mathbf{b}\lambda + \mathbf{a}\epsilon$. What this implies is that $\mathbf{c} \wedge \mathbf{a} \wedge \mathbf{b} = 0$.

[3] This chapter is an optional read. Do skip this if this is your first read.

and $\mathbf{\Lambda}$ is a C^k form field on N, then $d(z^*\mathbf{\Lambda}) = z^*d(\mathbf{\Lambda})$. If Σ is a function on N, the function $z^*\Sigma$ on M is defined by the mapping z as the function whose value at a point p on the manifold M is the value of Σ at the image of the point $z(p)$, i.e., $z^*\Sigma(p) = \Sigma(z(p))$. What this implies is that z^* maps function linearly from N to M, similar to how z maps points from M to N. Now, if $\zeta(t)$ is a curve existing in M and passing via the point p, then the image of this curve $z(\zeta(t))$ existing in N passes via the image point $z(p)$. Consider a tangent vector to the image curve, denoted by $z_*\left(\frac{\partial}{\partial t}\right)_\zeta|_p$. z_* acts as a linear map of the tangent space $T_p(M)$ into the tangent space $T_{z(p)}(N)$. For each C^r function $\mathbf{G}$ and vector $\mathbf{R}$ at $z(p)$, $R(z^*g(p)) = z^*R(f(z(p))$. We can make use of this mapping $z_* : M \to N$ to define a linear 1-form mapping as $z^* : T^*_{z(p)}(N) \to T^*_p(M)$. This implies that an arbitrary p-form $\mathbf{\Lambda} \in T^*_{z(p)}$ is mapped onto the p-form $z^*\mathbf{\Lambda} \in T^*_p$, such that $\langle z^*\mathbf{\Lambda}, \mathbf{R}\rangle_p = \langle \mathbf{\Lambda}, z_*\mathbf{R}\rangle_{z(p)}$ is true for a vector $R \in T_p$. The fact that $d(z^*\mathbf{\Lambda}) = z^*d(\mathbf{\Lambda})$ holds is a consequence of the previously mentioned result.

The exterior derivative acts on a function $\mathbf{g}$ which is merely a 0-form field to produce a 1-form field dg. Let's generalize this, let $\mathbf{\Lambda}$ be a p-form field defined by $\mathbf{\Lambda} = \Lambda_{\alpha\beta...\zeta}dx^\alpha \wedge dx^\beta ... \wedge dx^\zeta$. Now, take the exterior derivative of this p-form field to obtain a $(p+1)$-field as follows

$$d\mathbf{\Lambda} = d\Lambda_{\alpha\beta...\zeta}dx^\alpha \wedge dx^\beta ... \wedge dx^\zeta \tag{3.20}$$

This $(p+1)$-field is independent of the coordinates $\alpha, \beta, ..., \zeta = x^\alpha$ used in definition and to convince ourselves of this fact, consider another set of coordinates, say their barred counterparts given by $\bar{\alpha}, \bar{\beta}, ..., \bar{\zeta} = x^{\bar{\alpha}}$. In these coordinates we have $\mathbf{\Lambda} = \Lambda_{\bar{\alpha}\bar{\beta}...\bar{\zeta}}dx^{\bar{\alpha}} \wedge dx^{\bar{\beta}} ... \wedge dx^{\bar{\zeta}}$. The components of $\Lambda_{\bar{\alpha}\bar{\beta}...\bar{\zeta}}$ are given by

$$\Lambda_{\bar{\alpha}\bar{\beta}...\bar{\zeta}} = \frac{\partial x^\alpha}{\partial x^{\bar{\alpha}}}\frac{\partial x^\beta}{\partial x^{\bar{\beta}}}...\frac{\partial x^\zeta}{\partial x^{\bar{\zeta}}}\Lambda_{\alpha\beta...\zeta} \tag{3.21}$$

and the new definition of $d\mathbf{\Lambda}$ in the coordinates of $x^{\bar{\alpha}}$ is the following [4]

[4] The last line was arrived at as $\frac{\partial^2 x^\alpha}{\partial x^{\bar{\alpha}}\partial x^{\bar{\chi}}}$ is symmetric in $\bar{\alpha}$ and $\bar{\chi}$, but $dx^{\bar{\chi}} \wedge dx^{\bar{\alpha}}$ is skew symmetric.

$$
\begin{aligned}
d\mathbf{\Lambda} &= d\Lambda_{\bar{\alpha}\bar{\beta}...\bar{\zeta}} dx^{\bar{\alpha}} \wedge dx^{\bar{\beta}} ... \wedge dx^{\bar{\zeta}} \\
&= d\left(\frac{\partial x^{\alpha}}{\partial x^{\bar{\alpha}}}\frac{\partial x^{\beta}}{\partial x^{\bar{\beta}}}\cdots\frac{\partial x^{\zeta}}{\partial x^{\bar{\zeta}}}\Lambda_{\alpha\beta...\zeta}\right) dx^{\bar{\alpha}} \wedge dx^{\bar{\beta}} ... \wedge dx^{\bar{\zeta}} \\
&= \frac{\partial x^{\alpha}}{\partial x^{\bar{\alpha}}}\frac{\partial x^{\beta}}{\partial x^{\bar{\beta}}}\cdots\frac{\partial x^{\zeta}}{\partial x^{\bar{\zeta}}} d\Lambda_{\alpha\beta...\zeta} dx^{\bar{\alpha}} \wedge dx^{\bar{\beta}} ... \wedge dx^{\bar{\zeta}} \qquad (3.22) \\
&+ \frac{\partial^2 x^{\alpha}}{\partial x^{\bar{\alpha}}\partial x^{\bar{\chi}}}\frac{\partial x^{\beta}}{\partial x^{\bar{\beta}}}\cdots\frac{\partial x^{\zeta}}{\partial x^{\bar{\zeta}}} d\Lambda_{\alpha\beta...\zeta} dx^{\bar{\chi}} \wedge dx^{\bar{\alpha}} \wedge dx^{\bar{\beta}} ... \wedge dx^{\bar{\zeta}} \\
&= d\Lambda_{\alpha\beta...\zeta} dx^{\alpha} \wedge dx^{\beta} ... \wedge dx^{\zeta}
\end{aligned}
$$

Consider the coordinate expression for dg, $dg = \frac{\partial g}{\partial x^{\alpha}} dx^{\alpha}$, observe that $d(dg) = \frac{\partial^2 g}{\partial x^{\alpha}\partial x^{\beta}} dx^{\alpha} \wedge dx^{\beta} = 0$. This is due to [4]. This implies that for any p-form field $\mathbf{\Lambda}$, $d(d\mathbf{\Lambda}) = 0$.

3.10 General p-forms

The gradient of a scalar produces a 1-form and the exterior derivative of this 1-form produces a 2-form, this chain continues. Thus the exterior derivative of a $(p-1)$-form produces a p-form defined as follows

$$
\mathbf{\Xi} = \frac{1}{p!} \Xi_{i_1\ i_2\ ...\ ip} dx^{i_1} \wedge dx^{i_2} \wedge ... \wedge dx^{i_p} \qquad (3.23)
$$

and the exterior derivative of $\mathbf{\Xi}$ is defined as follows

$$
d\mathbf{\Xi} = \frac{1}{p!} \frac{\partial \Xi_{i_1\ i_2\ ...\ ip}}{\partial x^{i_0}} dx^{i_0} \wedge dx^{i_1} \wedge dx^{i_2} \wedge ... \wedge dx^{i_p} \qquad (3.24)
$$

It is important to note that $\Xi_{i_1\ i_2\ ...\ ip}$ is antisymmetric under a 2 index interchange. The definitions provided here is an alternate one to the one in which the factor of $\frac{1}{p!}$ is not included[5] A *closed form* is defined as a form whose differential is zero while an *exact form* is defined as a form that is the differential of an exterior form (and it's an example of a closed form).

[5] i.e., in accordance to this definition, a 2-form is written as $dx^{\alpha} \wedge dx^{\beta} = \frac{1}{2}\left(dx^{\alpha} \otimes dx^{\beta} - dx^{\beta} \otimes dx^{\alpha}\right)$.

3.11 Parallel Transport and Covariant Differenriation

An ordinary differential of a vector $\mathbf{A}^{\mu}$ in a direction x^{α} is defined as follows

$$\frac{\partial \mathbf{A}^{\mu}}{\partial x^{\alpha}} dx^{\alpha} = \partial_{\alpha} \mathbf{A}^{\mu} dx^{\alpha} = \mathbf{A}^{\mu}_{,\alpha} \equiv \mathbf{A}^{\mu}(x + dx) - \mathbf{A}^{\mu}(x) \tag{3.25}$$

The ordinary differentials are defined by the difference between two vectors defined at two distinct points. In curved spacetime, however, we need to account for the rotations undergone by the vector as it evolves with time. Thus, we introduce the quantity $\delta \mathbf{A}^{\mu}$ and subtract it from the ordinary differential to obtain the covariant differential $D_{\alpha} \mathbf{A}^{\mu} \equiv \mathbf{A}^{\mu}_{;\alpha}$. To observe this rotation due to curvature, we transport the vector $\mathbf{A}^{\mu}(x + dx)$ to the point x without changing it's direction (see figure). This is known as parallel transport.

$$\mathbf{A}^{\mu}_{;\alpha} \equiv \mathbf{A}^{\mu}(x + dx) - [\mathbf{A}^{\mu}(x) + \delta \mathbf{A}^{\mu}(x)] \tag{3.26}$$

Let $\mathbf{S}$ be a $\begin{pmatrix} 1 \\ 1 \end{pmatrix}$ tensor. The covariant derivative $D_{\mathbf{R}}\mathbf{S}$ of $\mathbf{S}$ along a curve $P(\zeta)$, whose tangent vector $\mathbf{R} = \frac{dP}{d\zeta}$ is defined as follows

$$D_{\mathbf{R}}\mathbf{S}|_{P(0)} = lt_{\eta \to 0} \left[\frac{\mathbf{S}[P(\eta)]_{II^{el}\ transported\ to P(0)} - \mathbf{S}[\mathbf{P(0)}]}{\eta} \right] \tag{3.27}$$

The covariant derivative, denoted by either ∇ or D is a *connection* at a point p on the manifold M which allots every vector field $\mathbf{R}$ at p a differential operator $D_{\mathbf{R}}$, such that the operator maps an arbitrary C^r vector field $\mathbf{S}$ into a vector field $D_{\mathbf{S}}$. Following are some of the algebraic properties of D.

1. $D_{\mathbf{R}}\mathbf{S}$ is a tensor in the argument $\mathbf{R}$. For arbitrary functions g, h and continuous, once-differentiable vectors fields, i.e., C^1 vectors fields $\mathbf{R}, \mathbf{S}, \mathbf{Q}$,

$$D_{g\mathbf{R}+h\mathbf{S}}\mathbf{Q} = gD_{\mathbf{R}}\mathbf{Q} + hD_{\mathbf{S}}\mathbf{Q} \tag{3.28}$$

2. $D_{\mathbf{R}}\mathbf{S}$ obeys the linearity condition.

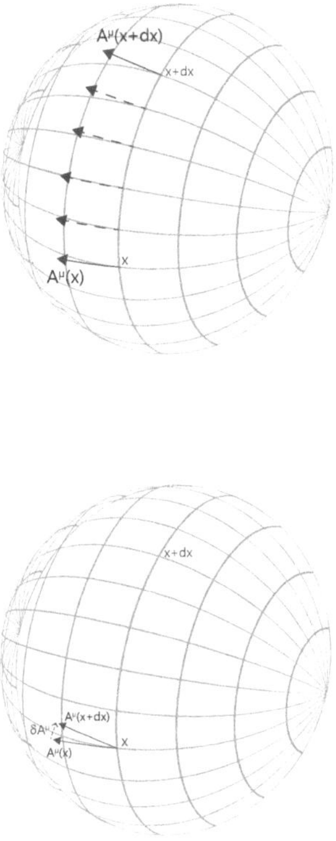

Fig. 3.3. The evolution of the vector along the curved path (L) and the parallel transport of $\mathbf{A}^{\mu}(x+dx)$ to the point x

$$D_{\mathbf{R}}\left(\mu\mathbf{S}+\nu\mathbf{Q}\right)=\mu D_{\mathbf{R}}\mathbf{S}+\nu D_{\mathbf{R}}\mathbf{Q} \tag{3.29}$$

3. For any two C^1 vector fields of the same rank $\mathbf{R},\mathbf{S}$

$$D_{\mathbf{R}}\mathbf{S}-D_{\mathbf{S}}\mathbf{R}=[\mathbf{R},\mathbf{S}] \tag{3.30}$$

In the third property, $[\mathbf{R},\mathbf{S}]$ is called a *commutator*. Suppose $\mathbf{R}$ and $\mathbf{S}$ are tangent vectors fields, then it holds that $\mathbf{R}=\partial_{\mathbf{R}}$ and

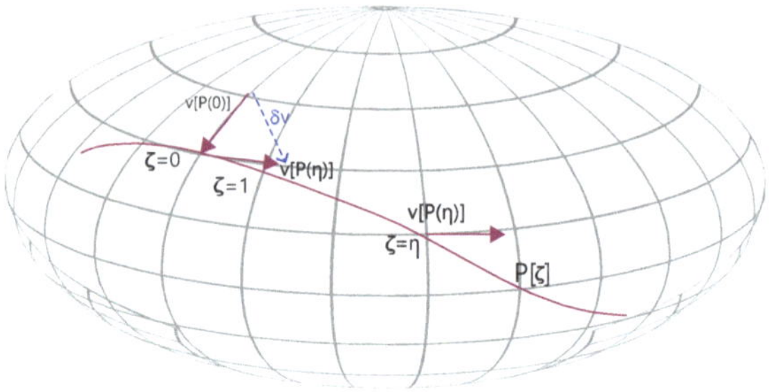

Fig. 3.4. Parallel transport of a vector on the parameterized curve $P(\zeta)$

$\mathbf{S} = \partial_{\mathbf{S}}$ are true (from previous definitions). Thus, a commutator, which by itself is a tangent vector field is defined as follows

$$[\mathbf{R}, \mathbf{S}] \equiv [\partial_{\mathbf{R}}, \partial_{\mathbf{S}}] \equiv \partial_{\mathbf{R}}\partial_{\mathbf{S}} - \partial_{\mathbf{S}}\partial_{\mathbf{R}} \tag{3.31}$$

Here, there is a need to define the commutation coefficients of a basis as the concept will come in handy for a future topic. For the basis vectors $\mathbf{e}_{\boldsymbol{\alpha}}$ and $\mathbf{e}_{\boldsymbol{\beta}}$, the commutation coefficient $C^{\gamma}_{\alpha\beta}$ is obtained by commuting the basis vectors. It is a tensor-like coefficient which gives the difference between partial derivatives of two coordinates with respect to the other coordinate.

$$[\mathbf{e}_{\boldsymbol{\alpha}}, \mathbf{e}_{\boldsymbol{\beta}}] \equiv \partial_{\alpha}\mathbf{e}_{\boldsymbol{\beta}} - \partial_{\beta}\mathbf{e}_{\boldsymbol{\alpha}} \equiv C^{\gamma}_{\alpha\beta}e_{\gamma} \tag{3.32}$$

If $C^{\gamma}_{\alpha\beta} = 0$ it is called a coordinate basis or *holonomic* if some $C^{\gamma}_{\alpha\beta}0$ then it is called a non-coordinate basis or *anholonomic*.

3.12 Connection Coefficients: An Introduction

From the example involving parallel transportation of the vector $\mathbf{A}^{\mu}$, for a small dx, $\delta\mathbf{A}^{\mu}(x)$ should be linear in dx and also in $\mathbf{A}^{\mu}(x)$ or in other words, it must be an output of some transformation of the vector $\mathbf{A}^{\mu}(x)$ at x. This is given below.

$$\delta\mathbf{A}^{\mu}(x) = B^{\mu}_{\nu}A^{\nu}(x) \tag{3.33}$$

Here, B^{μ}_{ν} is a matrix that transforms the vector during parallel transport. During parallel transport, the basis vectors and the basis 1-forms would twist, contract, expand, and turn according to the curvature, and this is quantified by the connection coefficient. The connection coefficient is defined as follows

$$\begin{aligned} \Gamma^{\gamma}_{\alpha\beta} &= \langle \boldsymbol{\omega}^{\gamma}, D_{\mathbf{e}_{\alpha}}\mathbf{e}_{\beta} \rangle = \langle \boldsymbol{\omega}^{\gamma}, D_{\alpha}\mathbf{e}_{\beta} \rangle \\ \Gamma^{\gamma}_{\alpha\beta} &= -\langle D_{\alpha}\boldsymbol{\omega}^{\gamma}, \mathbf{e}_{\beta} \rangle \end{aligned} \tag{3.34}$$

It is fairly easy to prove the latter equation (proof given below). The matrix mentioned previously is defined as $B^{\mu}_{\nu} = -\Gamma^{\mu}_{\nu\alpha}dx^{\alpha}$. Thus, we can conclude that the matrix accounts for all the contributions of the basis vectors and the basis 1-forms via the connection coefficients, over a small distance dx.

The Proof
To Prove that $\Gamma^{\gamma}_{\alpha\beta} = -\langle D_{\alpha}\boldsymbol{\omega}^{\gamma}, \mathbf{e}_{\beta} \rangle$

From lemma: $\langle \boldsymbol{\omega}^{\gamma}, \mathbf{e}_{\beta} \rangle = \delta^{\gamma}{}_{beta}$

$D_{\alpha}\langle \boldsymbol{\omega}^{\gamma}, \mathbf{e}_{\beta} \rangle = \partial_{\mathbf{e}_{\alpha}}\langle \boldsymbol{\omega}^{\gamma}, \mathbf{e}_{\beta} \rangle = D_{\alpha}\left(\delta^{\gamma}_{\beta}\right) = 0$

Thus, $0 = \underbrace{(D_{\alpha}\boldsymbol{\omega}^{\gamma}) \otimes \mathbf{e}_{\beta} + \boldsymbol{\omega}^{\gamma} \otimes (D_{\alpha}\mathbf{e}_{\beta})}_{D_{\alpha}(Contraction\ of\ \boldsymbol{\omega}^{\gamma}\otimes\mathbf{e}_{\beta})}$

$0 = \underbrace{\langle D_{\alpha}\boldsymbol{\omega}^{\gamma}, \mathbf{e}_{\beta} \rangle + \langle \boldsymbol{\omega}^{\gamma}, D_{\alpha}\mathbf{e}_{\beta} \rangle}_{Contraction\ of\ [D_{\alpha}(\boldsymbol{\omega}^{\gamma}\otimes\mathbf{e}_{\beta})]}$

$\Gamma^{\gamma}_{\alpha\beta} = -\langle D_{\alpha}\boldsymbol{\omega}^{\gamma}, \mathbf{e}_{\beta} \rangle$

When we take the covariant derivative of a tensor, we are to differentiate the tensor with respect to the arbitrary basis and also account for the twisting and turning of the 1-forms and the vectors present in the tensor's slots. Consider a tensor **S** of rank $\binom{1}{1}$. Upon covariant differentiation, we obtain the following terms

$$S^{\beta}_{\alpha;\gamma} = S^{\beta}_{\alpha,\gamma} + \Gamma^{\beta}_{\nu\gamma} S^{\nu}_{\alpha} - \Gamma^{\nu}_{\alpha\gamma} S^{\beta}_{\nu} \tag{3.35}$$

There are three things to note here which will help in understanding how to take the covariant derivative for any arbitrarily ranked tensor. Firstly, it is to be noted that a + (positive) sign is used if the index being corrected is upstairs. In the example, an arbitrary summation index ν was used to correct β that resided upstairs, i.e., $+\Gamma^{\beta}_{\nu\gamma} S^{\nu}_{\alpha}$. The second point to be noted is the use of a – (negative) sign. It is to be employed when the index being corrected is downstairs, $-\Gamma^{\nu}_{\alpha\gamma} S^{\beta}_{\nu}$. Lastly, we observe that the index being corrected shifts from the tensor **S** onto the connection Γ and is replaced on the tensor by a dummy summation index ν. Let's see a few examples to strengthen this concept. One way to check the correctness of your answer is to check for homogeneity- check if the indexes upstairs and downstairs are alike on either sides of the equation (this is shown in the examples).

1. $\underbrace{D_{\gamma} T_{\alpha\beta}}_{\{\frac{1}{\gamma\alpha\beta}\}} = \underbrace{\frac{\partial T_{\alpha\beta}}{\partial x^{\gamma}}}_{\{\frac{1}{\gamma\alpha\beta}\}} - \underbrace{\Gamma^{\xi}_{\gamma\alpha} T_{\beta\xi}}_{\{\frac{\xi}{\gamma\alpha\beta\xi}\}} - \underbrace{\Gamma^{\xi}_{\gamma\beta} T_{\alpha\xi}}_{\{\frac{\xi}{\gamma\alpha\beta\xi}\}}$
2. $\underbrace{D_{\gamma} T_{\alpha}}_{\{\frac{1}{\gamma\alpha}\}} = \underbrace{T_{\alpha,\gamma}}_{\{\frac{1}{\gamma\alpha}\}} - \underbrace{\Gamma^{\beta}_{\gamma\alpha} T_{\beta}}_{\{\frac{1}{\gamma\alpha}\}}$

3.13 Structure Coefficients

Let M be a manifold equipped with a topology. Consider the set of coordinates $\{x^{alpha}\} = (x^1, x^2, ..., x^n)$ in the chart z. Now, the coordinates follow a lemma which states the following: $\frac{\partial x^{\alpha}}{\partial x^{\beta}} = \delta^{\alpha}_{\beta}$, this lemma also implies that $d^2 x^{\alpha} = 0$. For moving frames, however, the differentials of the 1-forms $\boldsymbol{\Xi}$ do not vanish, i.e., the wedge product of two 1-forms do not yield a null result. This wedge product produces a 2-form given by[6]

[6] As mentioned before, in the alternate formalism the fraction $\frac{1}{2}$ is omitted.

$$d\Xi^m \equiv -\frac{1}{2}C^m_{ab}\Xi^a \wedge \Xi^b \tag{3.36}$$

where C^m_{ab} is called the structure coefficients of the frame. The structure coefficient C^m_{ab} is antisymmetric in a and b.

3.14 Riemannian Connection

A Riemannian connection Ω is defined for a pseudo-Riemannian metric **G**. It is a linear connection obeys two conditions, the covariant derivative of the metric is zero, and the second condition requires that the second covariant derivatives of scalar functions to commute. The second condition implies that the connection has *vanishing torsion.*

Theorem 3.3. *The following conditions determine the Riemannian connection* $\Omega^\gamma_{\alpha\beta}$

$$\partial_\gamma g_{\alpha\beta} - \Omega^\lambda_{\gamma\beta} g_{\alpha\lambda} - \Omega^\lambda_{\gamma\alpha} g_{\lambda\beta} = 0 \tag{3.37}$$

Let h *be a scalar function, the condition requires the following*

$$D_\gamma \partial_\alpha h - D_\alpha \partial_\gamma h = 0 \tag{3.38}$$

The Riemannian connection is defined as follows

$$\begin{aligned} \Omega^\gamma_{\alpha\beta} &\equiv \Gamma^\gamma_{\alpha\beta} + g^{\gamma\lambda}\bar{\Omega}_{\alpha\beta,\lambda} \\ \bar{\Omega}_{\alpha\beta,\lambda} &\equiv \tfrac{1}{2}\left(g_{\lambda\xi}C^\xi_{\alpha\beta} - g_{\xi\beta}C^\xi_{\alpha\lambda} - g_{\alpha\xi}C^\xi_{\beta\lambda}\right) \end{aligned} \tag{3.39}$$

a coordinate basis for which the structure coefficients ($C^\xi_{\alpha\beta}$ *and others) are zero, is called holonomic . A non-coordinate basis always has some non-zero structure coefficients, and is called anholonomic . In the holonomic case, the connection coefficients are called Christoffel symbols given by the following expression (will be proven later)*

$$\Gamma^\gamma_{\alpha\beta} \equiv \frac{1}{2}g^{\gamma\lambda}\left(g_{\beta\lambda,\alpha} + g_{\alpha\lambda,\beta} - g_{\alpha\beta,\lambda}\right) \tag{3.40}$$

3.15 Revisiting the Metric Tensor

Using the slot machine definition of tensors, we can think of the metric tensor as a slot machine with two slots which accept only vectors as inputs: $g(\underbrace{\quad}_{vector1}, \underbrace{\quad}_{vector2})$. When the same vector is inserted into the slots, we get the square of the length of the vector as the output, $g(\mathbf{R}, \mathbf{R}) = \mathbf{R}^2$. When two different vectors are inserted, we obtain the scalar product of the vectors as the output. It is important to note that irrespective of the order of insertion of the vectors, the result remains unchanged. This is shown below

$$g(\mathbf{R}, \mathbf{Q}) = g(\mathbf{Q}, \mathbf{R}) = \mathbf{RQ} = \mathbf{QR} \tag{3.41}$$

The metric obeys the condition of linearity and in a specific coordinate system, its operation on the two input vectors is given by the following bilinear expression

$$g(\mathbf{R}, \mathbf{Q}) = g_{\mu\nu} R^{\mu} Q^{\nu} \tag{3.42}$$

There exists a reason for the name metric tensor, at least for the case when the inner product is positive definite. Consider two points, z and $z + \Delta z$, infinitesimally close to each other. The square of the infinitesimal distance of the displacement vector with components Δz^m is represented as follows

$$(\Delta s)^2 \equiv g_{mn} \Delta z^m \Delta z^n \tag{3.43}$$

Now, the metric tensor, just like any other tensor, would transform under a coordinate change as follows

$$g_{mn}(x) = \frac{\partial \bar{x}^i}{\partial x^m} \frac{\partial \bar{x}^j}{\partial x^n} g_{ij}(\bar{x}) \tag{3.44}$$

from the above transformation, it is clear that the length Δs of the displacement vector is not dependent on the choice of coordinates, rather it is dependent only on the two points under consideration. This formula is nothing but the generalization of the Pythagorean theorem of Euclidean geometry, which states that

$$(\Delta s)^2 = \Delta x^2 + \Delta y^2 + \Delta z^2 \tag{3.45}$$

the emergence of the metric tensor is the starting point of Riemanninan geometry. Another interesting way to look at it, and one which we will be extensively using in future chapters, is to observe that the length of a curve $P[\zeta(t)]$ between two points $\zeta(t_1)$ and $\zeta(t_2)$ can be expressed as follows

$$s = \int_{t_1}^{t_2} \left[g_{mn} \frac{d\zeta^m}{dt} \frac{d\zeta^n}{dt} \right]^{\frac{1}{2}} dt \tag{3.46}$$

3.16 Normal Coordinates

Consider the curve $P(\zeta)$ with well-defined endpoints, say a and b, and let R^γ be the tangent vector. The tensor $\mathbf{S}$ is said to be parallelly transported along the curve $P(\zeta)$ if $\frac{D\mathbf{S}}{s\zeta} = 0$. The covariant derivative of the tangent vector can be expressed in the terms of the metric tensor as $D_\alpha \mathbf{R}^\gamma = g^{\mu\gamma} D_\alpha \mathbf{R}_\mu$. Now,

$$D_\alpha \mathbf{R}^\gamma = D_\alpha \left(g^{\mu\gamma} \mathbf{R}_\mu \right) = g^{\mu\gamma} D_\alpha \mathbf{R}_\mu + \mathbf{R}_\mu D_\alpha g^{\mu\gamma} \tag{3.47}$$

but, we know that $D_\alpha \mathbf{R}^\gamma = g^{\mu\gamma} D_\alpha \mathbf{R}_\mu$. Hence, we conclude that $\mathbf{R}_\mu D_\alpha g^{\mu\gamma}$. This is not just something that we obtained from lousy reasoning, observe the term carefully, you would realize that it is the very condition of that of parallel transport. What this means is that we want the inner product of two vector inputs, say $\mathbf{a}$ and $\mathbf{b}$, $g(\mathbf{a}, \mathbf{b}) = \mathbf{ab} = g_{\mu\nu} a^\mu b^\nu$ to remain constant under parallel transport along a curve with tangent R^γ. This gives rise to the following condition

$$R^\gamma D_\gamma \left(g_{\mu\nu} a^\mu b^\nu \right) = 0 \tag{3.48}$$

parallel transport requires $R^\gamma a^\mu b^\nu D_\gamma g_{\mu\nu}$ to be true for all R, a, b. The vanishing covariant metric derivative is not a consequence of using any connection, it's a condition that allows us to choose a specific connection $\Gamma^\rho_{\mu\nu}$. In principle, we could have connections for which $D_\gamma g_{\mu\nu} 0$, but we specifically require a connection for which this condition is true because we want a parallel transport operation which preserves angles and lengths. In the local frame, which is the reference frame in the vicinity of an arbitrary point x_0 in which we can choose normal coordinates[7] such that at that point $g_{\mu\nu}(x_0) = \delta_{\mu\nu}$ and the derivative of the metric with respect to any component of the metric can be set to 0, i.e., $g_{\mu\nu,\alpha} = 0$ and also such that $\frac{\partial g_{\mu\nu}}{\partial x^\mu \partial x^\nu} 0$ (except when space is flat). The last condition implies that at the local point x_0, the connection vanishes (specifically, the Christoffel symbol vanishes), i.e., $\Gamma^\rho_{\mu\nu} = 0$.

Consider the locally flat coordinates (or normal coordinates) $\xi^i(x^\mu)$, it can be shown that $\frac{\partial^2 \xi^1}{\partial x^\mu \partial x^\nu} = \Gamma^\rho_{\mu\nu} \frac{\partial \xi^1}{\partial x^\rho}$[8]. It can be shown by the following calculation that the covariant derivative of the metric tensor vanishes.

$$
\begin{aligned}
D_\rho g_{\mu\nu} &= \partial_\rho g_{\mu\nu} - g_{\mu\sigma}\Gamma^\sigma_{\nu\rho} - g_{\sigma\nu}\Gamma^\sigma_{\mu\rho} \\
&= \partial_\rho \left(\frac{\partial \xi^i}{\partial x^\mu} \frac{\partial \xi^i}{\partial x^\nu} \right) - g_{\mu\sigma} \frac{\partial x^\sigma}{\partial \xi^i} \frac{\partial^2 \xi^i}{\partial x^\nu \partial x^\rho} - g_{\sigma\nu} \frac{\partial x^\sigma}{\partial \xi^i} \frac{\partial^2 \xi^i}{\partial x^\mu \partial x^\rho} \\
&= \frac{\partial^2 \xi^i}{\partial x^\rho \partial x^\mu} \frac{\partial \xi^i}{\partial x^\nu} + \frac{\partial \xi^i}{\partial x^\mu} \frac{\partial^2 \xi^i}{\partial x^\rho \partial x^\nu} - \frac{\partial \xi^j}{\partial x^\mu} \underbrace{\frac{\partial \xi^j}{\partial x^\sigma} \frac{\partial x^\sigma}{\partial \xi^i}}_{\delta^j_i} \frac{\partial^2 \xi^i}{\partial x^\nu \partial x^\rho} - \frac{\partial \xi^j}{\partial x^\sigma} \frac{\partial \xi^j}{\partial x^\nu} \frac{\partial x^\sigma}{\partial \xi^i} \frac{\partial^2 \xi^i}{\partial x^\mu \partial x^\rho} \\
&= 0
\end{aligned}
\tag{3.49}
$$

3.17 Pfaffian Derivatives

coframe on a manifold M is a system of 1-forms which form a basis of the cotangent bundle at every point (just to remind ourselves- The dual space $^*T_p$ is often called the cotangent space). The system of 1-forms Ξ^m used in defining the structure coefficients is a coframe. The Pfaffian derivatives ∂_m in the coframe Ξ^m of a function h is defined as follows

$$dh \equiv \partial_m h \ \Xi^m \tag{3.50}$$

[7] sometimes called Gaussian normal coordinates

[8] This equation will be proven in future chapters and it has a very deep physical meaning.

it is important to note that Pfaffian derivatives, unlike normal derivatives, do not commute. This can be seen from the analysis of the identity $dh \equiv 0$ as follows

$$
\begin{aligned}
&d^2h \equiv \tfrac{1}{2}\left[\partial_m\partial_n h - \partial_n\partial_m h - C^a_{mn}\partial_a h\right]\ \Xi^m \wedge \Xi^n \equiv 0 \\
&\partial_m\partial_n h - \partial_n\partial_m h = C^a_{mn}\partial_a h
\end{aligned}
\tag{3.51}
$$

The basis $\mathbf{e}_m$ which is the dual to Ξ^m satisfies the commutation conditions (in this formalism we include the fraction $\frac{1}{2}$, however, for all future purposes we will neglect this factor)

$$
C^a_{mn}e_a = [\mathbf{e}_m, \mathbf{e}_n] \tag{3.52}
$$

3.18 Back to Connections

Consider a tensor $\mathbf{S}$ of rank $\begin{pmatrix}1\\1\end{pmatrix}$. With the knowledge of the components of $S^\beta_{\alpha;\gamma}$ we can calculate the components of the covariant derivative $D_{\mathbf{R}}\mathbf{S}$ by a contraction into R^γ as follows (where $\mathbf{R} = \frac{dP}{d\zeta} = \frac{dx^\gamma}{d\zeta}$ is a tangent vector present on the curve $P(\zeta)$)

$$
D_{\mathbf{R}}\mathbf{S} = \left(S^\beta_{\alpha;\gamma}R^\gamma\right)\mathbf{e}_\beta \otimes \boldsymbol{\omega}^\alpha \tag{3.53}
$$

The components of $D_{\mathbf{R}}\mathbf{S}$ are denoted by $\frac{DS^\alpha_\beta}{d\zeta}$. Thus, we obtain

$$
\begin{aligned}
&\frac{DS^\beta_\alpha}{d\zeta} \equiv S^\beta_{\alpha;\gamma}R^\gamma = S^\beta_{\alpha;\gamma}\frac{dx^\gamma}{d\zeta} \\
&\frac{DS^\beta_\alpha}{d\zeta} = \frac{dS^\beta_\alpha}{d\zeta} + \left(\Gamma^\beta_{\nu\gamma}S^\nu_\alpha - \Gamma^\nu_{\alpha\gamma}S^\beta_\nu\right)\frac{dx^\gamma}{d\zeta}
\end{aligned}
\tag{3.54}
$$

To find the connection coefficients for a given basis we first need to take metric coefficients in the given basis and then calculate their directional derivatives along the considered basis directions.

$$
\begin{aligned}
&D_\gamma g_{\alpha\beta} = g_{\alpha\beta,\gamma} - \Gamma^\nu_{\alpha\gamma}g_{\nu\beta} - \Gamma^\nu_{\beta\gamma}g_{\nu\alpha} = 0 \\
&g_{\alpha\beta,\gamma} - \Gamma_{\beta\alpha\gamma} - \Gamma_{\alpha\beta\gamma} = 0 \\
&g_{\alpha\beta,\gamma} - 2\Gamma_{(\alpha\beta)\gamma}
\end{aligned}
\tag{3.55}
$$

Let us now construct a metric for $\Gamma_{\nu\alpha\gamma}$.

$$\begin{aligned}\tfrac{1}{2}\left(g_{\nu\beta,\gamma}+g_{\nu\gamma,\beta}-g_{\beta\gamma,\nu}\right) &= \Gamma_{(\nu\beta)\gamma}+\Gamma_{(\nu\gamma)\beta}-\Gamma_{(\beta\gamma)\nu} \\ &= \tfrac{1}{2}\left(\Gamma_{\nu\beta\gamma}+\Gamma_{\beta\nu\gamma}+\Gamma_{\nu\gamma\beta}+\Gamma_{\gamma\nu\beta}-\Gamma_{\beta\gamma\nu}-\Gamma_{\gamma\beta\nu}\right) \quad (3.56) \\ &= \Gamma_{\nu\beta\gamma}+\left(\Gamma_{\beta[\nu\gamma]}+\Gamma_{\gamma[\nu\beta]}-\Gamma_{\nu[\beta\gamma]}\right)\end{aligned}$$

Let $(\mathbf{R}=\mathbf{e}_\nu, \mathbf{Q}=\mathbf{e}_\lambda)$ be two basis vectors. We can now use them to construct structure coefficients by commuting the basis which was something we realized in the chapter on Pfaffian derivatives.

$$\begin{aligned} [\mathbf{e}_\nu,\mathbf{e}_\lambda] &= D_\nu\mathbf{e}_\lambda - D_\lambda\mathbf{e}_\nu = C^\rho_{\nu\lambda}e_\rho \\ C^\rho_{\nu\lambda}e_\rho &= \left(\Gamma^\rho_{\lambda\nu}-\Gamma^\rho_{\nu\lambda}\right)e_\rho = 2\Gamma^\rho_{[\lambda\nu]}e_\rho \quad (3.57) \\ \Gamma^\rho_{[\lambda\nu]} &= -\tfrac{1}{2}C^\rho_{\nu\lambda} \quad \rightarrow \quad \Gamma_{\rho[\lambda\nu]} = -\tfrac{1}{2}C_{\nu\lambda\rho}\end{aligned}$$

Combining the equations we obtain an expression for the connection coefficient.

$$\Gamma_{\nu\beta\gamma}=\frac{1}{2}[g_{\nu\beta,\gamma}+g_{\nu\gamma,\beta}-g_{\beta\gamma,\nu}+\underbrace{C_{\nu\beta\gamma}+C_{\nu\gamma\beta}-C_{\beta\gamma\nu}}_{=\,0\ for\ coordinate\ basis\ (holonomic)}] \quad (3.58)$$

Thus, we obtain the Christoffel symbol which can be expressed as follows after raising an index.

$$\Gamma^\alpha_{\beta\gamma}=g^{\alpha\nu}\Gamma_{\nu\beta\gamma} \quad (3.59)$$

3.19 Transformation Formula for Connections

Let us take $\mathbf{S}=\mathbf{e}_\alpha=\frac{\partial}{\partial x^\alpha}$ to be the basis vector field (whose components are constants), and let $\mathbf{R}=\mathbf{e}_\beta=\frac{\partial}{\partial x^\beta}$. We can now expand $D_\mathbf{R}\mathbf{S}$ in the basis and the coefficients of expansion, $\Gamma^\rho_{\beta\alpha}$ is given below. This relation between $D_\mathbf{R}\mathbf{S}$ and $\Gamma^\rho_{\beta\alpha}$ is established here.

$$D_{\frac{\partial}{\partial x^\beta}}\left(\frac{\partial}{\partial x^\alpha}\right) = \Gamma^\rho_{\beta\alpha}\frac{\partial}{\partial x^\rho} \tag{3.60}$$

The transformation formula can now be derived using the above definition.

$$\begin{aligned}
D_{\frac{\partial}{\partial \bar{x}^\beta}}\left(\tfrac{\partial}{\partial \bar{x}^\alpha}\right) &= \bar{\Gamma}^\rho_{\beta\alpha}\tfrac{\partial}{\partial \bar{x}^\rho} \\
&= D_{\left(\frac{\partial x^\mu}{\partial \bar{x}^\beta}\right)\frac{\partial}{\partial x^\mu}}\left(\tfrac{\partial x^\gamma}{\partial \bar{x}^\alpha}\tfrac{\partial}{\partial x^\gamma}\right) \\
&= \tfrac{\partial x^\mu}{\partial \bar{x}^\beta} D_{\frac{\partial}{\partial x^\mu}}\left(\tfrac{\partial x^\gamma}{\partial \bar{x}^\alpha}\tfrac{\partial}{\partial x^\gamma}\right) \\
&= \tfrac{\partial x^\mu}{\partial \bar{x}^\beta}\left[\tfrac{\partial x^\gamma}{\partial \bar{x}^\alpha} D_{\frac{\partial}{\partial x^\mu}}\left(\tfrac{\partial}{\partial x^\gamma}\right) + \left(\tfrac{\partial}{\partial x^\mu}\tfrac{\partial x^\gamma}{\partial barx^\alpha}\right)\tfrac{\partial}{\partial x^\gamma}\right] \\
&= \left[\tfrac{\partial x^\mu}{\partial \bar{x}^\beta}\tfrac{\partial x^\gamma}{\partial \bar{x}^\alpha}\Gamma^\sigma_{\mu\gamma} + \left(\tfrac{\partial^2 x^\sigma}{\partial \bar{x}^\beta \partial \bar{x}^\alpha}\right)\right]\tfrac{\partial}{\partial x^\sigma}
\end{aligned} \tag{3.61}$$

where the dummy index γ is replaced with σ. To make the comparison between the last line and the first line of the derivation, we need to manipulate the partial factor $\frac{\partial}{\partial x^\sigma}$ as follows

$$\frac{\partial}{\partial x^\sigma} = \frac{\partial \bar{x}^\rho}{\partial x^\sigma}\frac{\partial}{\partial \bar{x}^\rho} \tag{3.62}$$

thus, we obtain

$$\begin{aligned}
\bar{\Gamma}^\rho_{\beta\alpha}\tfrac{\partial}{\partial \bar{x}^\rho} &= \left[\tfrac{\partial x^\mu}{\partial \bar{x}^\beta}\tfrac{\partial x^\gamma}{\partial \bar{x}^\alpha}\Gamma^\sigma_{\mu\gamma} + \left(\tfrac{\partial^2 x^\sigma}{\partial \bar{x}^\beta \partial \bar{x}^\alpha}\right)\right]\tfrac{\partial \bar{x}^\rho}{\partial x^\sigma}\tfrac{\partial}{\partial \bar{x}^\rho} \\
\bar{\Gamma}^\rho_{\beta\alpha}(\bar{x}) &= \tfrac{\partial \bar{x}^\rho}{\partial x^\sigma}\tfrac{\partial x^\mu}{\partial \bar{x}^\beta}\tfrac{\partial x^\gamma}{\partial \bar{x}^\alpha}\Gamma^\sigma_{\mu\gamma}(x) + \tfrac{\partial^2 x^\sigma}{\partial \bar{x}^\beta \partial \bar{x}^\alpha}\tfrac{\partial \bar{x}^\rho}{\partial x^\sigma}
\end{aligned} \tag{3.63}$$

The three index notation seems to suggest that the Christoffel symbol is a tensor of rank three. This, however, is not true, the proof is in the extra term that appears in the transformation above. Due to this very same confusion Christoffel symbols, in older notations, were written as $\left\{ {\rho \atop \beta\alpha} \right\}$ instead of $\Gamma^\rho_{\beta\alpha}$.

3.20 Torsion Tensor

The torsion tensor $T^\rho_{\alpha\beta}$ is a third-rank tensor, antisymmetric in the first two indices and with 24 independent components.

$$T^{\rho}_{\alpha\beta} \equiv \Gamma^{\rho}_{[\alpha\beta]} \tag{3.64}$$

Consider the transformation of the Christoffel symbol again, considering the antisymmetric part of the transformation we can show that the torsion tensor does transform like a third-rank tensor. This brings us to a very important conclusion, that the torsion tensor cannot be eliminated locally due the reason that if a tensor vanishes at a particular point then it vanishes everywhere.

$$\begin{aligned} \Gamma^{\rho}_{\alpha\beta}(x) &= \frac{\partial x^{\sigma}}{\partial \bar{x}^{\rho}} \frac{\partial \bar{x}^{\beta}}{\partial^{\mu}} \frac{\partial \bar{x}^{\alpha}}{\partial x^{\gamma}} \bar{\Gamma}^{\sigma}_{\mu\gamma} + \frac{\partial^2 x^{\sigma}}{\partial x^{\beta} \partial x^{\alpha}} \frac{\partial x^{\sigma}}{\partial \bar{x}^{\rho}} \\ \Gamma^{\rho}_{[\alpha\beta]} &= T^{\rho}_{\alpha\beta} = \bar{T}^{\sigma}_{\mu\gamma} \frac{\partial x^{\rho}}{\partial \bar{x}^{\sigma}} \frac{\partial \bar{x}^{\mu}}{\partial x^{\alpha}} \frac{\partial \bar{x}^{\gamma}}{\partial x^{\beta}} \end{aligned} \tag{3.65}$$

Let $\mathbf{R}$ and $\mathbf{S}$ be vectors. Without torsion $[\mathbf{R}, \mathbf{S}]$ and $D_{\mathbf{R}}\mathbf{S} - D_{\mathbf{S}}\mathbf{R}$ represent the same vector, i.e., $[\mathbf{R}, \mathbf{S}] = D_{\mathbf{R}}\mathbf{S} - D_{\mathbf{S}}\mathbf{R}$. $D_{\mathbf{R}}\mathbf{S} - D_{\mathbf{S}}\mathbf{R} - [\mathbf{R}, \mathbf{S}] = 0$ represents a closed loop, and in the presence of torsion, there is no closure of loop. This is shown below.

$$\mathbf{R}^{\alpha} D_{\alpha} \mathbf{S}^{\beta} - \mathbf{S}^{\alpha} D_{\alpha} \mathbf{R}^{\beta} - [\mathbf{R}, \mathbf{S}]^{\beta} = T^{\beta}_{\alpha\gamma} R^{\alpha} S^{\gamma} \tag{3.66}$$

This describes the geometrical meaning of torsion, which is that torsion represents the failure of the loop to close. For all future calculations, we shall assume torsion-free connections, i.e., $\mathbf{T} = 0$.

4

The Three Types of Vectors

4.1 Non-Degeneracy of a Metric

A metric is said to be *non-degenerate* at a point p on a manifold M if there exists no non-zero vector $\mathbf{R} \in T_p(M)$ such that $g(\mathbf{R}, \mathbf{Q}) = 0$ for all vectors $\mathbf{Q} \in T_p(M)$. We can now define a new metric tensor of rank $\begin{pmatrix} 2 \\ 0 \end{pmatrix}$ with components $g^{\mu\nu}$ with respect to a basis $\{x_\mu\}$ which is dual to the basis $\{x^\mu\}$, by the following expression

$$g^{\mu\nu} g_{\nu\xi} = \delta^{\mu}_{\xi} \tag{4.1}$$

the matrix $g^{\mu\nu}$ is the inverse of the matrix $g_{\mu\nu}$, and these tensors can be used to provide an isomorphism between any contravariant tensor and a covariant one, i.e., to raise and lower indices. If $\mathbf{S}^{\mu\nu}$ are the components of a contravariant tensor, then we can lower its indices by making use of metric tensors, and can also obtain mixed tensors as follows

$$\begin{aligned} S_{\mu\nu} &= g_{\mu\xi} g_{\nu\chi} S^{\xi\chi} \\ S^{\mu}_{\nu} &= g^{\mu\xi} S_{\xi\nu} \\ S^{\nu}_{\mu} &= g^{\nu\chi} S_{\chi\mu} \end{aligned} \tag{4.2}$$

4.2 Timelike, Spacelike and Lightlike Vectors

Consider a Lorentzian metric $\mathbf{g}$ on a manifold M equipped with some topology. At a point p on the manifold, the non-zero vectors can be divided into three classes.

Type	**Condition**
Timelike	$g(\mathbf{R}, \mathbf{R}) < 0$
Lightlike	$g(\mathbf{R}, \mathbf{R}) = 0$
Spacelike	$g(\mathbf{R}, \mathbf{R}) > 0$

Table 4.1. All the conditions are mentioned for a vector $\mathbf{R} \in T_p$

4.3 Null Cones

A vector $\mathbf{R} \in T_p$ is called *causal* if

$$g(\mathbf{R}, \mathbf{R}) \leq 0 \tag{4.3}$$

At each point p on a Lorentzian manifold M, we can define a double cone C_p in the tangent space $T_p(M)$. This is called the *causal cone* and is expressed in terms of the following inequality

$$g(\mathbf{R}, \mathbf{R}) \leq 0, \ \ \mathbf{R} \in T_p(M) \tag{4.4}$$

The boundary of the causal cone is called the *double cone.* The boundary is formed by the null or lightlike vectors in the tangent space of the Lorentzian manifold and this separates the timelike and the spacelike vectors.

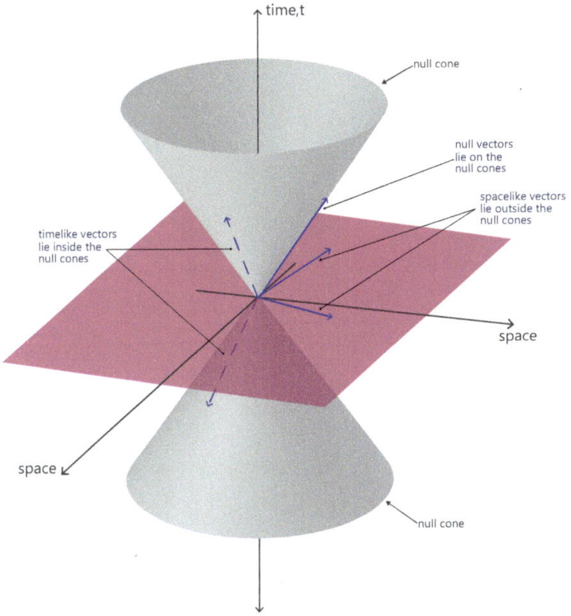

Fig. 4.1. This figure represents the null cones as defined by the Lorentz metric.

5

Geodesic equation

5.1 Introduction

A geodesic is defined as a spacetime curve that is the shortest distance between two points, straight and uniformly parameterized, or in other words, it is a curve whose distance between two points is stationary. Mathematically, a geodesic is a curve $P(\zeta)$ that parallel-transports its tangent vector, say $\mathbf{R} = \frac{dP}{d\zeta}$, along with itself.

$$D_{\mathbf{R}}\mathbf{R} = 0 \tag{5.1}$$

In a local coordinate system $x^{\eta}[P(\zeta)]$, in which the tangent vector takes the form $R^{\eta} = \frac{dx^{\eta}}{d\zeta}$, the geodesic is expressed as follows

$$\frac{D\left(\frac{dx^{\eta}}{d\zeta}\right)}{d\zeta} = 0 = \frac{d\left(\frac{dx^{\eta}}{d\zeta}\right)}{d\zeta} + \left[\Gamma^{\eta}_{\alpha\beta}\frac{dx^{\alpha}}{d\zeta}\right]\frac{dx^{\beta}}{d\zeta} \tag{5.2}$$

simplifying this gives us the geodesic equation.

$$\frac{d^2x^{\eta}}{d\zeta^2} + \Gamma^{\eta}_{\alpha\beta}\frac{dx^{\alpha}}{d\zeta}\frac{dx^{\beta}}{d\zeta} = 0 \tag{5.3}$$

5.2 Affine Parameter

If a geodesic is timelike,
1. It is a possible curve (or trajectory) for a freely falling observer,

and
2. there exists a parameter ζ (called the *affine parameter*) which is a multiple of the observer's propertime, $\zeta = m\tau + c$

5.3 The Deeper Meaning: Part One

Let us try and reveal the deeper meaning that hides in plain sight. In normal coordinates, in a local frame (for a local observer), we know that the following conditions are satisfied: $g_{\mu\nu,\alpha} = 0$ and $\Gamma^{\mu}_{\nu\beta} = 0$. The 4-velocity, which is the tangent vector of a timelike curve is defined as $\mathbf{u} = \frac{dx^\eta}{d\tau}\mathbf{e}_\eta|_{\eta=0} = \frac{dx^0}{d\tau}\mathbf{e}_0 = \mathbf{e}_0$. This is so because $\mathbf{u}$ and $\mathbf{e}_0$ both have unit length. Since the 4-velocity is constant, the 4-acceleration is zero, i.e.,

$$a = D_{\mathbf{u}}\mathbf{u} = D_0\mathbf{e}_0 = 0 \tag{5.4}$$

this equation is nothing but the previously defined geodesic equation. Comparing the equations we conclude the following

$$a = D_{\mathbf{u}}\mathbf{u} = D_0\mathbf{e}_0 = \Gamma^{\eta}_{00}\mathbf{e}_\eta = 0 \tag{5.5}$$

what this implies is that a freely falling observer experiences zero 4-acceleration, i.e., the observer moves along a geodesic with affine parameter equal to the observer's propertime. The geodesic equation for this observer in local coordinate is as follows

$$\frac{d^2x^\eta}{d\tau^2} + \Gamma^{\eta}_{\alpha\beta}\frac{dx^\alpha}{d\tau}\frac{dx^\beta}{d\tau} = 0 \tag{5.6}$$

Thus, we have proven that the observer's trajectory is a straight line![1]

5.4 The Deeper Meaning: Part Two

For the second reveal consider a particular spacelike coordinate, say $x^\eta = y$. Let the observer move slowly (i.e., at non-relativistic

[1] more generally, freely falling particles move on straight lines

speeds), this assumption enables us to replace propertime with just time. Let α and β (indices present in the geodesic equation) be timelike components, this would mean that $\frac{dx^\alpha}{d\tau} = \frac{dx^\beta}{d\tau} = \frac{dt}{T} = 1$[2], and $\Gamma^\eta_{\alpha\beta} = \Gamma^\eta_{tt}$. Making all these changes in the geodesic equation, we obtain the following expression

$$\frac{d^2y}{dt^2} = -\Gamma^\eta_{tt} \tag{5.7}$$

this looks freakishly similar to the expression for gravitational force. Well, hold on, let's dig a bit deeper.

At large distances from the spherically symmetric gravitating object, space-time is flat. Why? This is due to the fact that the influence of the gravitational field vanishes at large distances as it varies as $\approx r^{-2}$. Due to this, the ability of the tidal forces to curve space-time at large distances from the gravitating object fades away thus resulting in a flat space. Note that when we talk about the gravitating object, we have assumed that there is no matter in the surroundings of our object, thus the vacuum field. We can represent this flat space-time in terms of the line element as

$$ds^2 = dt^2 - \frac{1}{c^2}\left(dx^2 + dy^2 + dz^2\right) \tag{5.8}$$

now, how would this metric change in the vicinity of our object? For this let us first, write the basic form of the metric for a plane in polar coordinates. In flat space, the spatial distance between two points on a plane in polar coordinates is given by the following equation

$$ds^2 = r^2 d\theta^2 + dr^2 \tag{5.9}$$

now, let us change this metric. We first start by making the replacements, $sin \rightarrow sinh\omega$, and $cos \rightarrow cosh\omega$. Here, ω is the angle with which the hyperbola increases with respect to the origin (a timelike coordinate). Thus, we have changed from polar coordinates to *hyperbolic coordinates*. In this frame, the acceleration along a particular hyperbola is the same, however, the acceleration along different

[2] although time and propertime are distinct conceptually, they are dimensionally the same quantity and thus cancel

hyperbolae are different. An analogous relation can be drawn to that of circular motion here, similar to the acceleration remaining the same along a particular hyperbola, the acceleration of a particle moving around a circle is uniform, however, the acceleration around another concentric circle of a different radius who definitely not be the same. We now make the transforms: $X = rcosh\omega$, $T = rsinh\omega$, such that

$$X^2 - T^2 = r^2 \left[cosh^2\omega - sinh^2\omega\right] = r^2 \tag{5.10}$$

hence, producing the metric

$$ds^2 = r^2 d\omega^2 - dr^2 \tag{5.11}$$

this is the metric in which our gravitating object lies, let us travel along a particular hyperbola and to determine the fate of the metric. Let the gravitating object, under consideration, be the supermassive black hole located at the centre of our galaxy. Now, let us remove all the matter present outside this black hole and for the moment assume that the value of the energy density of the vacuum of space is zero[3]. By performing these actions, we have established the vacuum conditions. From a small distance from a hyperbola that is present next to where Earth was, just a moment ago, we compute the metric. The black hole is almost 26,000 light years away from Earth. Placing the origin at the centre of the black hole, we re-define the position vector to be

$$r = R_{BH \to Hyp} + r' \tag{5.12}$$

Where $R_{BH \to Hyp} = 26,000$ light years, is the distance between the black hole and the hyperbola trajectory which runs next to where Earth was a moment ago, and r' is the distance between the hyperbola and us. This distance is prone to vary since were nothing but mere particles floating in space but would never exceed that of $R_{BH \to Hyp}$, hence, $\frac{r'^2}{R^2_{BH \to Hyp}} \to 0$. Let us substitute this new relation into the metric and perform some manipulations.

---

[3] these conditions correspond to the Ricci flatness conditions in which we assume the stress-energy tensor and the cosmological constant to have a null value.

$$ds^2 = (R_{BH\to Hyp} + r')^2 \, d\omega^2 - [d(R_{BH\to Hyp} + r')]^2$$

$$ds^2 = \left(R^2_{BH\to Hyp} + r'^2 + 2R_{BH\to Hyp}r'\right) d\omega^2 - [dr']^2 \tag{5.13}$$

$$ds^2 \approx \left(1 + \frac{2r'}{R_{BH\to Hyp}}\right) R^2_{BH\to Hyp} d\omega^2 - dr'^2$$

We know that proper acceleration A, when the speed of light is set to unity is nothing but $\frac{1}{R}$ [4]. Hence, here, $A = \frac{1}{R_{BH\to Hyp}} = \mathbf{g}$ [5]. Define $R_{BH\to Hyp}\omega = t$.

$$ds^2 = (1 + 2r'\mathbf{g})\, dt^2 - dr'^2 = \left(1 + 2\frac{r'}{R}\right) dt^2 - dr'^2 \tag{5.14}$$

Now, we know that the Christoffel symbol is nothing but a combination of the partial derivatives of the metric tensor, with the assumptions made int the initial part of this chapter, it takes the following form

$$\Gamma^{\eta}_{tt} = \frac{1}{2} g^{\eta\gamma} \left(g_{t\gamma,t} + g_{\gamma t,t} - g_{tt,\gamma}\right) \tag{5.15}$$

from the metric (for the motion along the y-coordinate), $ds^2 = Adt^2 - dy^2$, $g^{y\gamma} = g^{yy} = 1$ (since η and γ assume a spacelike coordinate). Thus,

$$\Gamma^{y}_{tt} = \frac{1}{2} \left(g_{ty,t} + g_{yt,t} - g_{tt,y}\right) = -\frac{1}{2}\frac{\partial g_{tt}}{\partial y} \tag{5.16}$$

substituting this in the geodesic equation, we obtain an expression that in fact suggests the Christoffel symbol Γ^{y}_{tt} to be the force and g_{tt} to be the potential.

$$\frac{d^2y}{dt^2} = -\Gamma^{y}_{tt} = \frac{1}{2}\frac{\partial g_{tt}}{\partial y} \tag{5.17}$$

Replacing r' with y in the metric we obtained previously, we make an observation that confirms the statement made about the Christoffel symbol.

---

[4] $A = \frac{c}{T} = \frac{c}{R\times c}|_{c\equiv 1} = \frac{1}{R}$

[5] this $\mathbf{g}$ here is the acceleration due to gravity, not to be confused with the metric tensor.

$$
\begin{aligned}
ds^2 &= \underbrace{(1 + 2y\mathbf{g})}_{=-g_{tt}} dt^2 - dy^2 \\
-g_{tt} &= (1 + 2y\mathbf{g}) \quad \rightarrow \quad \frac{\partial g_{tt}}{\partial y} = -2\mathbf{g} \\
\frac{d^2 y}{dt^2} &= -\Gamma^y_{tt} = -\mathbf{g}
\end{aligned}
\tag{5.18}
$$

Thus, the equation of motion of a geodesic in an accelerated coordinate frames (for non-relativistic speeds) is nothing but Newton's equation in a uniform gravitational field.

6

Curvature

A straight line has zero curvature and a circle of radius ρ has a curvature $\frac{1}{\rho}$, well, why? A more important question is not why but how, how is the curvature defined for surfaces and based on what is it defined. The curvature of a curve is defined by how swiftly its unit normal vector $\mathbf{n}$ evolves as we move along the curve. For a small distance traversed on a circle, say ds, the infinitesimal change in the unit normal $|d\mathbf{n}|$ is equal to the angle $\frac{ds}{\rho}$. Similarly, the curvature of a straight line is zero because the normals are all parallel to each other and do not evolve with time. Hence, curvature is measured by the ratio of the infinitesimal change in the unit normal $|d\mathbf{n}|$ to the infinitesimal distance traversed by a point .

6.1 Gauss Curvature and Geodesic Deviation

Let S be a 2-dimensional surface with $\mathbf{n}(\xi)$ as unit normal. Let the local coordinates on this surface be denoted by $x(\xi) = (\xi^1, \xi^2)$ and the infinitesimal changes as a point traverse a distance be $d\xi^1, d\xi^2$. The infinitesimal distance traversed is then given by the expression given below which is tangential to the 2-dimensional surface.

$$dx = \frac{\partial x}{\partial \xi^1} d\xi^1 + \frac{\partial x}{\partial \xi^2} d\xi^2 \equiv x_{,\alpha} d\xi^\alpha \tag{6.1}$$

The unit normal is a vector that does not evolve as the point traverses because it is constant length $\mathbf{n.n} = 0$. This implies that the infinitesimal changes in it due to the parameters ξ are orthogonal to it. i.e., $d\mathbf{n.n} = 2\mathbf{n}d\mathbf{n} = 0$. Hence, we can conclude that

$d\mathbf{n} = \mathbf{n}_{,\alpha} d\xi^\alpha$ is tangential to the surface. When the tangential vectors $d\mathbf{n}_1$ and $d\mathbf{n}_2$ are expanded in the basis vector we obtain the following expression

$$\mathbf{n}_{,\alpha} = L^\beta_\alpha x_{,\beta} \tag{6.2}$$

the coefficients L^β_α define a mapping of the tangent vector dx into $d\mathbf{n}$, another tangent vector. The matrix L^β_α is called the *Weingarten matrix* and the mapping is called *Weingarten mapping.* This was the idea of curvature given by Carl Friedrich Gauss, which states that the curvature of a surface at a point is measured by the ratio of the area spanned by the infinitesimal components of the normal vector $d\mathbf{n}_1, d\mathbf{n}_2$, to the area spanned by the tangent vectors $dx_1 = x_{,1} d\xi^1$ and $dx_2 = x_{,2} d\xi^2$ on the surface.

Let $\mathbf{R}$ and $\mathbf{S}$ be two vectors, the area of the parallelogram spanned by these vectors is given as $|\mathbf{R} \times \mathbf{S}|$. The curvature is measured as follows

$$\begin{aligned} d\mathbf{n}_1 \times d\mathbf{n}_2 &= (\mathbf{n}_{,1} \times \mathbf{n}_{,2})\, d\xi^1 d\xi^2 \\ &= \left(L^1_1 x_1 + L^2_1 x_{,2}\right) \times \left(L^1_2 x_1 + L^2_2 x_{,2}\right) d\xi^1 d\xi^2 \\ &= \left(L^1_1 L^2_2 - L_1 L^2 L^1_2\right)(dx_1 \times dx_2) = |L|\,(dx_1 \times dx_2) \end{aligned} \tag{6.3}$$

where $|L| = \left(L^1_1 L^2_2 - L_1 L^2 L^1_2\right)$ is called Gauss curvature . Hence, the curvature of a plane is zero since the normals are all in the same direction. Similarly, the curvature of a cylinder is zero too because the normal does not change in a direction parallel to the axis of the cylinder although it changes along the circular surface. The curvature of a sphere, however, is $\frac{1}{\rho^2}$ (where ρ is the radius of the sphere) because the solid angle $d\Omega$ that it subtends at the sphere's center is the same as that spanned by the normals.

Consider two geodesics that are initially parallel, separated by a distance χ_0. They are no longer parallel when we traverse a distance s and their separation is measured by $\chi = \chi_0 cos\left(\frac{s}{\rho}\right)$, where ρ is the radius of the sphere. The separation follows a surprising equation.

$$\chi = \chi_0 cos\left(\frac{s}{\rho}\right)$$

$$\frac{d\chi}{ds} = -\frac{\chi_0}{\rho} sin\left(\frac{s}{\rho}\right) \tag{6.4}$$

$$\frac{d^2\chi}{ds^2} = -\frac{\chi_0}{\rho^2} cos\left(\frac{s}{\rho}\right) = -\frac{1}{\rho^2}\chi$$

$$\frac{d^2\chi}{ds^2} + R\chi = 0$$

Where $R = \frac{1}{\rho^2}$ is the Gaussian curvature of the surface and the equation derived is called the equation of geodesic deviation which is nothing but the equation of simple harmonic motion!

6.2 Theorema Egregium

Carl Friedrich Gauss found out that the curvature of a surface can be measured exclusively in terms of quantities intrinsic to the surface, without any reference to how the surface is embedded in the surrounding (3-dimensional) space. The intrinsic quantities are the coefficients $g_{\mu\nu}$ of the line element ds which measures the distance between two close points present on the surface. Consider a pair of points separated by an infinitesimal distance $d\xi^\alpha$, (ξ^1, ξ^2) and $(\xi^1 + d\xi^1, \xi^2 + d\xi^2)$. The separation between the points is expressed as follows

$$ds^2 = x_{,\mu} x_{,\nu} d\xi^\mu d\xi^\nu \equiv g_{\mu\nu} d\xi^\mu d\xi^\nu \tag{6.5}$$

this quadratic form of the metric is positive definite, symmetric and is non-singular (i.e., it's determinant is non-zero $|g|0$). The theorem states that it is the combination of the Weingarten matrices L^μ_ν that determines the Gauss curvature (also called total curvature), which is given in terms of the metric in an equation called the *Gauss equation.*

$$L^\mu_\nu L^\kappa_\alpha - L^\mu_\alpha L^\kappa_\nu = -g^{\mu\zeta} R^\kappa_{\zeta\nu\alpha} \tag{6.6}$$

$R^\kappa_{\zeta\nu\alpha}$ is called the *Riemann curvature tensor.* The straightest possible curves on this surface, i.e., the geodesics, are expressed as a function of $\xi^\alpha(s)$, where s is the distance measured along the curve, which satisfy the geodesic equation, $\frac{d^2\xi^\kappa}{ds^2} + \Gamma^\kappa_{\zeta\nu}\frac{d\xi^\zeta}{ds}\frac{d\xi^\nu}{ds} = 0$.

6.3 The Riemann Curvature Tensor

The Riemann curvature tensor is a higher-dimensional analogue of the Gaussian curvature. In 2-dimensions, the direction of the acceleration of one geodesic relative to another geodesic (called the *fiducial geodesic*) is fixed uniquely by the demand that their separation vector $\boldsymbol{\chi}$ is perpendicular to the fiducial geodesic. However, in higher dimensions, the separation vector does not only remain perpendicular to the fiducial geodesic but also rotates about it. In the slot machine analogy, the Riemann tensor is a machine that has three slots and in a coordinate system, the components can be written as a *trilinear function* (it obeys linearity).

$$\mathbf{r} = R(\mathbf{a}, \mathbf{b}, \mathbf{c}) \quad \rightarrow \quad r^\alpha = R^\alpha_{\beta\lambda\xi} a^\beta b^\lambda c^\xi \tag{6.7}$$

The equation of geodesic deviation in higher dimensions replaces the Gaussian curvature with the Riemannian curvature tensor and the spatial distance with the propertime. Let the unit tangent vector or the 4-velocity be $u^\alpha = \frac{dx^\alpha}{d\tau}$, then the equation of geodesic deviation is expressed as

$$\begin{gathered} \frac{D^2\chi}{d\tau^2} + R(\mathbf{u}, \chi, \mathbf{u}) = 0 \\ \frac{D^2\chi^\alpha}{d\tau^2} + R^\alpha_{\beta\lambda\xi}\frac{dx^\beta}{d\tau}\chi^\lambda\frac{dx^\xi}{d\tau} = 0 \end{gathered} \tag{6.8}$$

Thus, the Riemann tensor is an exterior 2-form taking values in the set of linear maps from the tangent plane to itself. The non-commutativity of covariant derivatives is a geometrical property of the metric. The commutation $(D_\alpha D_\beta - D_\beta D_\alpha)\, u^\mu$ of two covariant derivatives of a vector u is a mixed tensor with coefficients $R_{\alpha\beta}{}^{\mu}{}_{\nu}$ such that

$$(D_\alpha D_\beta - D_\beta D_\alpha)\, u^\mu = [D_\alpha, D_\beta]\, u^\mu = R_{\alpha\beta}{}^{\mu}{}_{\nu} u^\nu \tag{6.9}$$

The components of the Riemann tensor in a coordinate basis is defined below and the proof follows.

$$R^{\alpha}_{\beta\lambda\xi} = \langle \boldsymbol{\omega}^{\alpha}, [D_{\lambda}, D_{\xi}]\, \mathbf{e}_{\beta} \rangle$$
$$= \Gamma^{\alpha}_{\beta\xi,\lambda} - \Gamma^{\alpha}_{\beta\lambda,\xi} + \Gamma^{\alpha}_{\mu\lambda}\Gamma^{\mu}_{\beta\xi} - \Gamma^{\alpha}_{\mu\xi}\Gamma^{\mu}_{\beta\lambda} \tag{6.10}$$

The Proof

$R^{\mu}_{\beta\lambda\xi} = \langle \boldsymbol{\omega}^{\alpha}, [D_{\lambda}, D_{\xi}]\, \mathbf{e}_{\beta} \rangle = \langle \boldsymbol{\omega}^{\alpha}, (D_{\lambda}D_{\xi} - D_{\xi}D_{\lambda})\, \mathbf{e}_{\beta} \rangle$

$= \langle \boldsymbol{\omega}^{\alpha}, (D_{\lambda}(D_{\xi}\mathbf{e}_{\beta}) - D_{\xi}(D_{\lambda}\mathbf{e}_{\beta})) \rangle$

$= \left\langle \boldsymbol{\omega}^{\alpha}, \left(D_{\lambda}\left(\mathbf{e}_{\mu}\Gamma^{\mu}_{\beta\xi}\right) - D_{\xi}\left(\mathbf{e}_{\mu}\Gamma^{\mu}_{\beta\lambda}\right)\right) \right\rangle$

$= \left\langle \boldsymbol{\omega}^{\alpha}, \mathbf{e}_{\mu}\Gamma^{\mu}_{\beta\xi,\lambda} + \Gamma^{\mu}_{\beta\xi}(D_{\lambda}\mathbf{e}_{\mu}) - \mathbf{e}_{\mu}\Gamma^{\mu}_{\beta\lambda,\xi} - \Gamma^{\mu}_{\beta\lambda}(D_{\xi}\mathbf{e}_{\mu}) \right\rangle$

$= \left\langle \boldsymbol{\omega}^{\alpha}, \mathbf{e}_{\mu}\Gamma^{\mu}_{\beta\xi,\lambda} + \left(\mathbf{e}_{\nu}\Gamma^{\nu}_{\mu\lambda}\right)\Gamma^{\mu}_{\beta\xi} - \mathbf{e}_{\mu}\Gamma^{\mu}_{\beta\lambda,\xi} - \left(\mathbf{e}_{\nu}\Gamma^{\nu}_{\mu\xi}\right)\Gamma^{\mu}_{\beta\lambda} \right\rangle$

$= \left\langle \boldsymbol{\omega}^{\alpha}, \mathbf{e}_{\mu}\left(\Gamma^{\mu}_{\beta\xi,\lambda} - \Gamma^{\mu}_{\beta\lambda,\xi}\right) + \mathbf{e}_{\nu}\left(\Gamma^{\nu}_{\mu\lambda}\Gamma^{\mu}_{\beta\xi} - \Gamma^{\nu}_{\mu\xi}\Gamma^{\mu}_{\beta\lambda}\right) \right\rangle$

$= \left(\Gamma^{\mu}_{\beta\xi,\lambda} - \Gamma^{\mu}_{\beta\lambda,\xi}\right) \underbrace{\langle \boldsymbol{\omega}^{\alpha}, \mathbf{e}_{\mu} \rangle}_{=\delta^{\alpha}_{\mu}=1 \ if \ \alpha=\mu} + \left(\Gamma^{\nu}_{\mu\lambda}\Gamma^{\mu}_{\beta\xi} - \Gamma^{\nu}_{\mu\xi}\Gamma^{\mu}_{\beta\lambda}\right) \underbrace{\langle \boldsymbol{\omega}^{\alpha}, \boldsymbol{\nu} \rangle}_{=\delta^{\alpha}_{\nu}=1 \ if \ \alpha=\nu}$

Hence, $R^{\mu}_{\beta\lambda\xi} = \Gamma^{\mu}_{\beta\xi,\lambda} - \Gamma^{\mu}_{\beta\lambda,\xi} + \Gamma^{\nu}_{\mu\lambda}\Gamma^{\mu}_{\beta\xi} - \Gamma^{\nu}_{\mu\xi}\Gamma^{\mu}_{\beta\lambda}$
The Riemann curvature tensor is closely related to tidal forces, it represents the tidal force experienced by a particle moving along a geodesic.

6.4 Symmetries of the Riemann Tensor

The Riemann curvature tensor has, in 4-dimensions, $4 \times 4 \times 4 \times 4 = 256$ independent components. Observations reveal a variety of algebraic symmetries such as the first skew symmetry, the second skew symmetry, and the block symmetry.

$$R_{\alpha\beta\mu\nu} = R_{[\alpha\beta][\mu\nu]}, \quad R_{[\alpha\beta\mu\nu]} = 0, \quad R_{\alpha[\beta\mu\nu]} = 0. \tag{6.11}$$

All of the above symmetries reduce the Riemann tensor from 256 components to 20 independent components. The antisymmetry of $(\alpha\beta)$ and $(\mu\nu)$ in $R_{\alpha\beta\mu\nu}$ implies that there are $P = \frac{1}{2}n(n-1)$ different ways of choosing non-trivial pairs $(\alpha\beta)$ and P ways of choosing non-trivial pairs $(\mu\nu)$. This is classified as the first class which involve elements only of the type $R_{\alpha\beta\alpha\beta}$ which have $\frac{1}{2}n(n-1)$ different members due to antisymmetry in α and β. The second class involves elements with three distinct indices of type $R_{\alpha\mu\beta\mu}$, where μ can take n values and for each of these values the indices α, β can take two distinct values from the remaining $n-1$ possibilities[1]. Thus, there are $\frac{1}{2}n(n-1)(n-2)$ possibilities. The last algebraic symmetry, called the cyclic symmetry can be written alternatively as

$$R_{\alpha[\beta\mu\nu]} = \tfrac{1}{3!}\left(R_{\alpha\beta\mu\nu} - R_{\alpha\beta\nu\mu} + R_{\alpha\nu\beta\mu} - R_{\alpha\nu\beta\mu} + R_{\alpha\mu\neq\beta} - R_{\alpha\mu\beta\nu}\right)$$

$$R_{\alpha[\beta\mu\nu]} = \tfrac{1}{3}\left(R_{\alpha\beta\mu\nu} + R_{\alpha\nu\beta\mu} + R_{\alpha\mu\nu\beta}\right) = \tfrac{1}{3}\Delta_{\alpha\beta\mu\nu} = 0 \tag{6.12}$$

where the pair symmetries guarantee that $\Delta_{\alpha\beta\mu\nu}$ is totally antisymmetric such that $\Delta_{\alpha\beta\mu\nu} = 0$ is trivial unless all the indices are distinct. This cyclic property in the last three indices is trivial for the first and the second class and does not reduce the number of elements. The third class comprises of elements in which all the indices are different. The indices can be arranged as $R_{\alpha\beta\mu\nu}$ with $\alpha < \beta, \mu < \nu, \alpha < \mu$ due to the fact that all other variants can be related to these via symmetries. Now, there are $\frac{1}{2}n(n-1)$ pairs with $\alpha < \beta$ and $\frac{1}{2}(n-2)(n-3)$ pairs for $\mu < \nu$. Thus, there are $\frac{n(n-1)(n-2)(n-3)}{2}$ variants which are to be reduced by a factor of $\frac{1}{2}$ (halved) to only retain the elements with $\alpha < \mu$ and since each cyclic relation permits one out of the three elements to be expressed in terms of the other two, we can further reduce the number of independent elements by a factor of $\frac{2}{3}$. The number of added constraints are then the number of combinations of 4 indices that can be taken from n indices is

$$\frac{2}{3}\frac{1}{2}\frac{n(n-1)(n-2)(n-3)}{2} = 2\frac{n!}{(n-4)!4!} \tag{6.13}$$

The total number of independent components is then given by

$$\frac{n(n-1)}{2} + \frac{n(n-1)(n-2)}{2} + 2\frac{n!}{(n-4)!4!} = \frac{n^2\left(n^2-1\right)}{12} \tag{6.14}$$

In 4-dimensions, we then have $\frac{4^2\left(4^2-1\right)}{12} = 20$ independent com-

[1] we limit to $\alpha < \beta$ because the elements with $\alpha > \beta$ are related by an interchange of pairs $\alpha\mu$ and $\beta\mu$

ponents. $\Delta_{\alpha\beta\mu\nu}$ mentioned in the previous equations is called *Bianchi's first identity.* Besides the algebraic symmetries, there exist differential symmetries called *Bianchi identities* are given by the following expression

$$R^{\alpha}_{\beta[\xi\mu;\nu]} = 0 \tag{6.15}$$

The contraction of the Riemann tensor is called the *Ricci tensor*[2], $R_{\mu\nu} = R^{\alpha}_{\beta\mu\nu} g^{\beta}_{\alpha}$ and the contraction of the Ricci tensor is the *curvature scalar*, $R = R_{\mu\nu} g^{\mu\nu}$.

$$R_{\mu\nu} = \partial_\rho \Gamma^{\rho}_{\nu\mu} - \partial_\nu \Gamma^{\rho}_{\rho\mu} + \Gamma^{\rho}_{\rho\lambda} \Gamma^{\lambda}_{\nu\mu} - \Gamma^{\rho}_{\nu\lambda} \Gamma^{\lambda}_{\rho\mu} \tag{6.16}$$

The Ricci tensor is symmetric, i.e., $R_{\mu\nu} = R_{\nu\mu}$, and out of the $\frac{1}{2}n^2\left(n^2-1\right)$ algebraically independent components of the Riemann tensor $\frac{1}{2}n\left(n+1\right)$ of them can be represented by the components of the Ricci tensor. For $n = 1$, $R_{\alpha\beta\nu} = 0$; for $n = 2$, there exists only one independent component of the Riemann tensor, which is the curvature scalar; for $n = 3$, the Ricci tensor completely determines the curvature tensor.

6.5 Weyl Tensor

For values of $n > 3$ in $\frac{1}{2}n^2\left(n^2-1\right)$, the components of the Riemann curvature tensor apart from its own components are represented by the Weyl tensor $W_{\alpha\beta\mu\nu}$.

$$W_{\alpha\beta\mu\nu} = R_{\alpha\beta\mu\nu} + \frac{2}{n-2}\left(g_{\alpha[\nu}R_{\mu]\beta} + g_{\beta[\mu}R_{\nu]\alpha}\right) + \frac{2}{(n-1)(n-2)} R g_{\alpha[\mu} g_{\nu]\beta} \tag{6.17}$$

The Weyl tensor also possesses all three algebraic symmetries and in addition it can be thought of as that part of the curvature tensor such that all contractions vanish, i.e., $W^{\alpha}_{\beta\alpha\nu} = 0$. The Weyl tensor is a measure of the curvature of spacetime or, more generally, a pseudo-Riemannian manifold. It can be shown that the Weyl tensor of a three-dimensional pseudo-Riemannian manifold (M, g) is

[2] can be defined as a covariant, contravariant, or a mixed tensor.

identically zero. Like the Riemann curvature tensor, it expresses the tidal force that a body feels when moving along a geodesic. It differs from the Riemann curvature tensor in that it does not convey information on how the volume of the body changes, but rather only how the shape of the body is distorted by the tidal force.

The Weyl tensor is equal to the Riemann tensor in a Ricci-flat space ($R_{\mu\nu} = 0$). It is considered that the Weyl tensor embodies in some sense the non-Newtonian properties of the gravitational field, in particular, its radiation properties. This point of view is supported by the fact that the equations for massless fields, at least in four spacetime dimensions, are *conformally invariant* (this is a concept that will be explained soon). Since W is obviously zero for a flat metric, it is also zero if the metric is conformal to a flat metric. It can be proved that if $n > 3$, then the identical vanishing of the Weyl tensor implies that the metric is locally conformally flat. Consider two metrics, $\mathbf{G}$ and $\bar{g}$. These metrics are said to be conformal if and only if

$$\bar{g} = \omega^2 g \tag{6.18}$$

where ω is a non-zero differentiable function. If this condition is satisfied, then for any vectors $\mathbf{R}, \mathbf{Q}, \mathbf{S}, \mathbf{V}$ at a point p on the manifold M,

$$\frac{g(\mathbf{R}, \mathbf{Q})}{g(\mathbf{S}, \mathbf{V})} = \frac{\bar{g}(\mathbf{R}, \mathbf{Q})}{\bar{g}(\mathbf{S}, \mathbf{V})} \tag{6.19}$$

so angles and lightlike world-lines are preserved under conformal transformations. The null cone structure in the tangent space $T_p(M)$ is preserved by conformal transformations since for a vector $\mathbf{R} \in M$,

$$g(\mathbf{R}, \mathbf{R}) > 0, = 0, < 0 \textit{ implies } \bar{g}(\mathbf{R}, \mathbf{R}) > 0, = 0, < 0 \tag{6.20}$$

The metric components are related as follows

$$\bar{g}_{\mu\nu} = \omega^2 g_{\mu\nu} \tag{6.21}$$

This concept of conformal factors and how helps it to select a relevant two-dimensional part of a spacetime and to make its stereographic projection on a compact space is studied in *Penrose-Carter*

diagrams . The idea behind these diagrams is that under conformal maps (when the conformal factor is dropped) lightlike or null world lines and angles between them do not change.
Let $\bar{g}_{\mu\nu}$ be the metric tensor on a Riemannain space, we choose another metric such that

$$\bar{g}_{\mu\nu} = e^{2\Xi} g_{\mu\nu} \tag{6.22}$$

$g_{\mu\nu}$ differs by a positive factor at each point. These metrics are conformally related, and even if one of them is flat the other one is called conformally flat. The Weyl tensors, for the given metrics, are related as follows

$$\bar{W}_{\alpha\beta\mu\nu} = e^{2\Xi} W_{\alpha\beta\mu\nu} \tag{6.23}$$

Revisiting the cases for different values of n, we conclude the following. In the case of $n = 1$, it is implied there are zero truly independent components in the metric. The implication of the case when $n = 2$, is that in any 2-dimensional Riemann manifold it is a standard result that locally we can always choose coordinates to make the metric conformally at. The implication of the case when $n = 3$, is that in any 3-dimensional Riemann manifold it is a standard result that locally we can always choose coordinates to make the metric diagonal, i.e., $g_{mn} = Diag\,(g_{11}, g_{22}, g_{33})$ (all the non-diagonal elements are zero), i.e., Riemann 3-manifolds have metrics that is always locally *diagonalizable.*

7

Einstein's Field Equations

7.1 Newton v Einstein: The Missing Sun

Before we divulge into the details and study the mind-numbing field equations, let us take a step back to put things into perspective. We must now ask ourselves an important question, why a different theory? I mean we are all cool with Newton's stuff, and as if his theories weren't enough we are moving towards a much more complicated one. We are in a very tricky situation now, upon studying the field equations one may either give up complaining that the math is just too much or one may ignore the math for a moment and focus on understanding the elegance of the equations. Of course, my aim is to try and stimulate the latter. I strongly believe that in order to understand a theory born out of the power of sheer imagination it is our responsibility to try and appreciate it using our own. Personally, I have always been fascinated by two things- gravity and serious people getting drunk, and I have since, well ever, tried to understand the former by using every ounce of imagination I possess; as for the latter, I guess it's open to interpretation...

Let's first perform a *Gedankenerfahrung*, which might possibly explain the need for a new theory. Imagine that for the moment both Newton and Einstein are alive (of course, they are in every physicist's heart!) and that they are participants in a debate hosted by Wolfgang Pauli (of course this weird choice comes with a reason). Since it's obvious that there would arise tensions (strictly egotistical) in a room of physicists, the argument the two would have is almost inevitable. Let's not take sides as of yet and try and review what each of them has to say. In Newton's version, he states that there is a potential (let's call it ϕ) everywhere in space and it varies from place to place. This variation of the potential, or better, the differential variation of the potential in space gives rise to a field, i.e., $\mathbf{E} = \nabla\phi\,(r)$ (I'll let you know what ∇ stands for later). The field instructs the particles how to move and decides their acceleration, i.e., $\mathbf{F} = m\mathbf{a} = -m\nabla\phi\,(r)$. Hence, we obtain the following relation

$$\mathbf{a} = -\nabla\phi\,(r) \tag{7.1}$$

now, what equation would instruct the field and tell it how to behave? We are to find the differential change in the field in order to predict its characteristics and the differential change in space (volume differentiation) is known as divergence and is represented as follows

$$\nabla\mathbf{E} = \left(\frac{\partial}{\partial x}\hat{i} + \frac{\partial}{\partial y}\hat{j} + \frac{\partial}{\partial z}\hat{k}\right)E \tag{7.2}$$

We are to bear in mind that the field E, unlike the potential ϕ, is a vector quantity, and in order solve for the same we make use of *Gauss theorem* which states that

$$\int \nabla\mathbf{E}\;dxdydz = \int d\nu\;\mathbf{E}_n \tag{7.3}$$

Here ν is the surface area and $\mathbf{E}_n$ is the normal vector to the surface area as depicted in the figure. In one dimension we can write

$\int d\nu\;\mathbf{E}_n = -\frac{Gm}{r^2}\int d\nu = -\frac{Gm}{r^2}4\pi r^2 = -4\pi Gm$

$\int \nabla\mathbf{E}\;d^3x = \int d\nu\;\mathbf{E}_n = -4\pi Gm$

Thus,

$$\nabla\mathbf{E} = -4\pi G\left(\frac{\Delta m}{\Delta V}\right) = -4\pi G\rho = \nabla^2\phi \tag{7.4}$$

Mass instructs the field how to curve via the density, ρ, this equation is known as the *Poisson equation.* Thus, Newton would argue that since his theory makes sense, if the Sun were to suddenly vanish, then the Earth would immediately be flung in a direction

tangential to its orbit almost analogous to how a stone attached to a string would fly in a tangential path when released from rotations about an axis. After this rather elegant conclusion, Pauli would be quite convinced with the reasoning and one would require a counter of epic proportions in order to even sound sane. In Einstein's version, he starts off by stating that Newton is simply wrong and that he is ignoring a very important concept. Einstein, being the controversial avant-garde physicist he is, states that it is well known that light takes time to travel from the Sun and reach Earth, and since the Sun is approximately 149.27 million kilometres away from us, it would take close to 8.3 minutes for a ray of light to reach us from the Sun (Thus the Sun you observe while reading this book is what it was 8.3 minutes ago, furthermore, all that we see around us is the past!). He goes on to argue that when light itself (the fastest thing known to us) takes time to completely its journey, how can gravity be any faster? Thus, if the Sun were to suddenly go, we would get to know the same only after 8.3 minutes, and some $8.3+x$ minutes later, the Earth would fly off tangentially. Thus, Pauli would announce Einstein the winner and let Newton know that he was **not even wrong**!
In General Relativity, the equation: $\mathbf{a} = -\nabla\phi\,(r)$, is replaced by a statement. The statement tells us that once we gain knowledge of the geometry (i.e. g_{00} here), the rule is that particles move on spacetime geodesics. It is quite interesting how this ceases to be true.

7.2 Stress-Energy Tensor: The Messenger of Mass

Space tells mass how to move and mass tells space how to curve. Prior to observing the curvature, we are to probe for a quantity that will enable us to understand how much mass-energy is present in a unit volume. This quantity is the stress-energy tensor. Spacetime possesses multiple contributions of 4-momentum from all sorts of particles from different fields. The contributions also pour in from the electromagnetic fields, neutrino fields, etc. Thus, we can view spacetime as an ocean of 4-momentum and the flow of water in the ocean is described by the stress-energy tensor $\mathbf{T}$. Since $\mathbf{T}$ is a tensor, it has a slot machine definition. The stress-energy tensor program is a linear, and symmetric slot machine which accepts two vector inputs, i.e., $T(\underbrace{\ \ }_{IP_1},\underbrace{\ \ }_{IP_2})$. The output, for a given input, of **T** are as follows

Input a 4-velocity **u** of an object and leave the other space sans any input. This produces the output

$$T(\ ,\mathbf{u}) = T(\mathbf{u},\ ) = -\left\{\frac{dp}{dV}\right\}$$

Here, the RHS denotes the density of 4-momentum, i.e., the 4-momentum per unit volume as measured in the object's local Lorentz frame. In the component form, we have the following expression for the object with 4-velocity u^μ,

$$T^\mu_\nu u^\nu = T^\mu_\nu v^\nu = -\left(\frac{dp^\mu}{dV}\right)$$

Now, enter the 4-velocity of the object as the second input and enter any arbitrary unit vector **n** as the second input. The program displays the following output

$$T(\mathbf{u},\mathbf{n}) = T(\mathbf{n},\mathbf{u}) = -n\frac{dp}{dV}$$

Here, the RHS denotes the component of the 4-momentum density, as measured in the object's Lorentz frame. In the component form, we have the following expression,

$$T_{\mu\nu}v^\mu n^\nu = T_{\mu\nu}n^\mu v^\nu = -n_\alpha\frac{dp_\alpha}{dV}$$

Enter the 4-velocity of the object for either of the inputs.

$$T(v,v) = \{\text{density of mass} - \text{energy measured in his Lorentz frame}\}$$

Now select two spacelike basis vectors for the object, in it's Lorentz frame, $\mathbf{e}_i$ and $\mathbf{e}_j$. Input the basis vectors to the tensor program **T**. The output is the i,j component of the stress as measured by the object (well us in the place of the object), it can be expressed as follows

$$T_{ij} = T(\mathbf{e}_i,\mathbf{e}_j) = T_{ji} = T(\mathbf{e}_j,\mathbf{e}_i)$$

$$= \left\{\begin{matrix}\text{i} - \text{component of force acting from side } \text{x}^\text{j} - \delta \text{ to side } \text{x}^\text{j} + \\ \delta,\text{ across a unit surface area with perpendicular direction } \mathbf{e}_\text{j}\end{matrix}\right\}$$

$$= \left\{ \begin{array}{l} \mathrm{j-component\ of\ force\ acting\ from\ side\ x^i} - \delta \mathrm{\ to\ side\ x^i}+ \\ \delta,\ \mathrm{across\ a\ unit\ surface\ area\ with\ perpendicular\ direction\ } \mathbf{e}_\mathrm{i} \end{array} \right\}$$

Now you since know how to construct the stress-energy tensor for an object, lets probe further into its physical significance. A stress-energy tensor $T^{\alpha\beta}$ is the flux of the $\alpha\ \hat{th}$ component of 4-momentum across a surface of constant x^β, thus:

$$T^{\mu 0}$$

T^{00}: The flux of the 0th component of 4-momentum across the time surface,i.e., indicates the density of energy.

$T^{k0} = T^{0k}$: Energy flux across the surface at constant x^k, i.e., indicates the flow of energy along x^k.

$T^{kd} = T^{dk}$: Flux of k-momentum across d-surface, i.e., indicates stress.

7.3 Conservation: The Opulent Origin of The Field Equations

Consider the room you are in (assuming only a single door and I request you to imagine one if you are outside) filled with an arbitrary charge density, the dynamics and the fate of this charge density and it's current density (if it chooses to become due to motion out of the door) is described using the 4-current tensor J^α. It has four components, with the first one indicating charge density and the rest three indicating current density. With this picture in mind how does one define charge conservation? If you are to conserve a particular charge, does it mean that you draw a boundary over the distribution and prohibit it from moving? Or does it mean that you transfer that specific charge density into an imaginary box of finite volume and move it to infinity?

If we follow the first definition, then we are to still deal with the charge density present in the room whenever we do physics but pretend as though it doesn't exist (leading to an awkward situation). The second case seems legit, right? Nah, not really cause if we move the charge density to infinity it would mean that we are moving the charge box over a time interval, leading to the creation of a current and as it passes via different areas it takes the form of

Fig. 7.1. A visual representation of the analogy presented in this chapter. The figure (top) shows the presence of charge density in the room and the bottom figure shows the current as charge moves towards the door.

current density (leading to a messy situation). So, is the fate of conservation bound to be awkward or messy? The answer is neither, we are to change our perspective a bit here. Let's start viewing the charge density from the perspective of the field its present in. Let's define the field and *connect it* to the source (which is the charge here) in such a way that the conservation of the source shall be an automatic consequence of some condition imposed on the field. Assume a hypercube in the 4-D spacetime to be the volume element in which an event occurs (or simply the 4-dimensional extension of the room you are in). There is a mathematical theorem that states that the two-dimensional boundary of the three-dimensional boundary of the 4-dimensional cube is zero.

Rather than divulging into the mathematics, let's try and understand this intuitively- we have previously observed the unique working of the exterior derivative, let me remind you, a 1-form α which is a gradient $\alpha = df$, must satisfy $d\alpha = 0$ (since if α is a 1-form then f is a 0-form because its exterior derivative is equal to α; now, $d\alpha = ddf$, which is zero). It is to be noted that not all 1-forms follow this relation. If a 1-form does α satisfies $d\alpha = 0$, then it follows that *locally* it has the form $\alpha = df$ for some f. This is an instance of the *Poincaré lemma*, which says that if a p-form γ satisfies $d\gamma = 0$, then *locally* γ has the form $\gamma = d\epsilon$, for some $(p-1)$-form ϵ. Now, consider a f-form β in a coordinate patch, with coordinates $x^1, x^2, x^3, \ldots,\ x^n$, there exists an asymmetrical set of components $\beta_{p\ldots u} (= \beta_{[p\ldots u]})$ to represent β. This representation can be expressed as follows

$$\beta = \sum \beta_{p\ldots u} dx^p \wedge \ldots \wedge dx^u \tag{7.5}$$

The exterior derivative of a p-form is a $(p+1)$-form that is written $d\beta$, and which has components

$$(d\beta)_{qp\ldots u} = \frac{\partial}{\partial x^{[q}} \beta_{p\ldots u]} \tag{7.6}$$

Yup, the notation is all messed up due to the antisymmetrization which extends overall $(p+1)$ indices, including the one on the derivative symbol. Thus, we can formulate the *fundamental theorem of exterior calculus* . For a p-form ζ

$$\int_A d\zeta = \int_{\partial A} \zeta \tag{7.7}$$

Here A is some *compact* $(p+1)$-dimensional region whose oriented p-dimensional boundary (which is also compact) is ∂A. What is going on here? Well, let's see if we can understand the physical meaning of the integral and try and do something with it. First off, the meaning of *compact* here is a region with a specific property that any infinite sequence of points lying in A must *accumulate* at some arbitrary point that exists within A. An *accumulation point* z has a specific property associated to it that every open set in A which contains z, must also contain members of an infinite sequence, such that the points of the sequence get closer and closer to the point z, without any limit. Examples of compact surfaces include the surfaces of a 2-sphere and that of a torus. However, the Euclidean plane is a non-compact surface. We now move on to the other term, *oriented*, which refers to the allotment of a consistent sign convention at every point of A. For a 0-manifold, this orientation allots a positive or a negative sign to each point. For a 1-manifold, this orientation associates a direction to the curve via a symbol (arrow). For a 2-manifold, this orientation is the circulation of the tangent vector at each point. A great example for a non-orientable surface is the *Mobius strip*. Thus, the boundary ∂A of a compact oriented $(p+1)$-dimensional region A consists of those points of A that do not lie within itself. If A is well-behaved, then ∂A is a compact oriented p-dimensional region (which might be possibly empty). It's boundary $\partial\partial A$ is empty (and thus $\partial\partial = \partial^2 = 0$, which makes sense because we know that $dd = d^2 = 0$). Examples of this phenomenon include the boundary of the closed unit disc in the complex plane is the unit circle; the boundary of the 2-sphere is empty, etc.

Similarly, taking the example of a cube in 3-dimensions—the boundary of a cube is its faces (2-dimensional), and the boundary of the each of the faces are composed of four edges (1-dimensional), and all edges are used up in uniting one face to another (i.e. no edges are left out). We can conclude that the 1-dimensional boundary of the 2-dimensional boundary of the 3-dimensional cube is identically zero, i.e., the boundary of a boundary vanishes. We can extend this concept to a hypercube and state that the 2-dimensional boundary of a 3-dimensional boundary of a 4-dimensional cube is identically zero. From Maxwell's equations (presented here without proof), it is known that $d\ {}^*F = 4\pi\ {}^*J$, describes the features of the field ${}^*F$. Here, the equations are expressed in a coordinate-free geometric form where **F** is called the *Faraday tensor* is a mathematical object that describes the electromagnetic field in spacetime, and $^*F$ is called the dual of the tensor[1]. The Faraday tensor is associated with the antisymmetric matrix of six electromagnetic fields as follows (note the difference between **F** and $^*\mathbf{F}$)

$$F^{\mu\nu} = \begin{pmatrix} 0 & -E_x & -E_y & -E_z \\ E_x & 0 & B_z & -B_y \\ E_y & -B_z & 0 & B_x \\ E_z & B_y & -B_x & 0 \end{pmatrix}$$
$$^*F^{\mu\nu} = \begin{pmatrix} 0 & -E_z & +E_y & +E_x \\ +E_z & 0 & -E_x & -B_y \\ -E_y & -B_z & 0 & -B_z \\ +B_x & +B_y & +B_z & 0 \end{pmatrix} \tag{7.8}$$

Observing the above matrices we can arrive at a possible equation for the relation between **F** and $^*\mathbf{F}$ as

$$^*F_{\alpha\beta} = \frac{1}{2}\epsilon_{\mu\nu\alpha\beta}F^{\mu\nu}, \tag{7.9}$$

where $\epsilon_{\mu\nu\alpha\beta}$ is called the *Levi-Civita* symbol in 4-dimensions. It is defined as follows

$$\epsilon_{\mu\nu\alpha\beta} = \left\{ \begin{cases} +1, \ if \ (\mu,\nu,\alpha,\beta) is \ an \ even \ permutation \ of \ (1,2,3,4), \\ -1, \ if \ (\mu,\nu,\alpha,\beta) is \ an \ odd \ permutation \ of \ (1,2,3,4), \\ 0, \ otherwise \end{cases} \right\} \tag{7.10}$$

This form of Maxwell's equation is useful as it contains, within itself, the electrostatic and the electromagnetic equations, i.e., $(\nabla\mathbf{E} = 4\pi\rho)\,,\left(\frac{\partial\mathbf{E}}{\partial t} - \nabla\times\mathbf{B} = -4\pi\mathbf{J}\right)$ respectively. Observe that the 4-current tensor contains the elements of the RHS of either equations, i.e., $J^\alpha = \left(J^0, J^1, J^2, J^3\right) = (\rho, \{J\})$ and as explained previously, since the Faraday tensor describes the electromagnetic field, it must hence contain the electric and magnetic fields as its components. Thus, a single geometric law is used to express the two Maxwell's equations as follows

[1] this is not to be confused with the dual lemma presented in introductory chapters. Here, the meaning of dual is associated to the *Hodge star operator*.

$$F^{\mu\nu}_{,\nu} = 4\pi J^{\mu} \tag{7.11}$$

an equivalent formalism of the coordinate-independent law would be to express the equation as $\nabla\mathbf{F} = 4\pi\mathbf{J}$. The conservation of the source, i.e., $d^*J = 0$, is a direct consequence of the identity $dd^*F = 0$. Thus, conservation is a direct consequence of the vanishing of a boundary of a boundary. Conservation literally demands no creation or destruction of the source inside the 4-dimensional cube. It is also to be noted that the integral of an event leading to a creation, i.e. of d^*J, over this 4-dimensional region is to be zero. Thus, conservation means zero creation of charge in a 4-dimensional region. Mathematically speaking, the application of the exterior derivative to either side of the second Maxwell equation, i.e., $d\ {}^*F = 4\pi\ {}^*J$, and using the fact that $d^2 = 0$, we can deduce that the 4-current $\mathbf{J}$ satisfies the vanishing boundary condition $d^*J = 0$, or $\nabla_\alpha J^\alpha = 0$ since $dd^*F = 4\pi d^*J = 0$ is true. This vanishing divergence of the 4-current yields a conservation law for electric charge. From the fundamental theorem of exterior calculus, we can write the conservation law as follows

$$\int_A d^*J = \int_{\partial A} {}^*J = 0 \tag{7.12}$$

the same law expressed in the *boundary of a boundary vanishes* language is the following

$$4\pi \int_{\partial A} {}^*J = \int_{\partial A} d^*F = \int_{\partial\partial A} {}^*F = 0 \tag{7.13}$$

So it is possible to use a similar reasoning to prove that there exist no magnetic monopoles? Indeed, we can; its Maxwell's show all the way. Magnetic charge is linked with the electromagnetic field via the equation: $4\pi J_{magnetic} = dF$. Thus, if any magnetic charge is not present then it would imply that the integral of $J_{magnetic}$ over any 3-volume A is zero (the fundamental theorem of exterior calculus); or

$$\int_A dF = \int_{\partial A} F = \{\text{magnetic flux passing } \textit{via } \partial A\} = 0 \tag{7.14}$$

and as mentioned previously this can be expressed in terms of the *boundary of a boundary vanishes* language by introducing a

4-potential, V (called vector potential). Thus, we express the Faraday tensor as $F = dV$ (V is a 1-form and its exterior derivative produces the 2-form, F), and have (this automatically leads to conservation, something that we wanted when we started off this topic)

$$\int_{\partial A} F = \int_{\partial A} dV = \int_{\partial\partial A} V \equiv 0 \tag{7.15}$$

thus, this concept of the *boundary of a boundary vanishes* is utilized to extend to the concept of conservation of either of Maxwell's equations, i.e., $dF = 0$, and $d^*J = 0$. Now, it is only natural to think if such a conservation is valid in gravitational physics and if it leads us to laws. Yup! It's true, conservation is a key concept in gravitational physics too but here we make use of a so-called *double dual* $^*R^*$ of the Riemann tensor which has the following relation with the Einstein tensor $\mathbf{G}$ and the stress-energy tensor $\mathbf{T}$

$$\mathbf{G} = Tr^*R^* = 8\pi\mathbf{T} \tag{7.16}$$

where Tr is the trace of the matrix (of the Riemann double dual tensor here). The reason for the 8π would become clear once we derive the field equations. The conservation of the source here is expressed as $d^*T = 0$, and it's a consequence of $d^*G = 0$, which is called *the contracted Bianchi identity.* However, unlike what the meaning of the vanishing boundary meant for charge, the meaning of the same in gravitational physics is expressed via net moment of rotation of a hypercube[2]. Conservation of the stress-energy tensor for a hypercube can be expressed by making use of the fundamental theorem of exterior as follows

$$\int_{A} d^*T = \int_{\partial A} {}^*T = 0 \tag{7.17}$$

7.4 Conservation Leads to Continuity?

Imagine there exists a charge distribution in the room you are presently in (if you're outside then assume an imaginary box around

---

[2] reference: Misner, C. W., Thorne, K. S., and Wheeler, J. A. (2017). Gravitation. Princeton University Press.

the size of your room), now suppose you decide to send the charge off to some distant place, then as it leaves your room, the charge flows as a current through the walls of the room (since a moving charge generates a current). Not only does the charge generate a current, but also generates a current density as the current flows out, say the door of your room, i.e., across a particular area. This leads to the idea of *continuity.*

Let the amount of charge present in your room (i.e., charge per unit volume of your room or charge density) be $\rho \left(= \frac{q}{A_{room}}\right)$. As the charge passes via the walls, the amount of charge leaving the room per unit time is $-\dot{\rho}(t)$ (negative sign indicates the fact that it's leaving your reference frame, i.e., your room). Now, you already know that this leaving charge is travelling through the walls in the form of current (current density actually since charge density is travelling not charge), thus giving rise to a changing current in the x, y, & z coordinates, i.e., giving rise to a diverging current density which can be expressed as ∇J. This relation can be expressed as follows

$$\begin{aligned} -\dot{\rho}(t) &= \left(\tfrac{\partial}{\partial x} + \tfrac{\partial}{\partial y} + \tfrac{\partial}{\partial z}\right) J = \nabla J, \\ \dot{\rho}(t) &+ \textstyle\sum_{\alpha=1}^{3} \tfrac{\partial J^\alpha}{\partial x^\alpha} = 0 \end{aligned} \tag{7.18}$$

we already know that $\mathbf{J}$ is the 4-current, and its components are $\left(J^0, J^1, J^2, J^3\right) = (\rho, J_x, J_y, J_z)$, hence we can express the above equation as follows

$$\frac{\partial J^\alpha}{\partial x^\alpha} = 0, \tag{7.19}$$

this is the continuity equation. This form of the continuity is valid in your room which is situated in flat space but in curved spacetime, continuity takes a different form. The difference is (quite obvious)- replace the partial derivative with a covariant derivative to take the form

$$\frac{DJ^\alpha}{\partial x^\alpha} = J^\alpha_{;\alpha} = 0 \tag{7.20}$$

What is to be noted here is that electric charge by itself is an invariant, i.e., it does not change no matter how it moves. The same cannot be said for charge and current density because they are components of a 4-current vector. A similar continuity equation can be proposed for gravitational physics in terms of the stress-energy tensor as follows

$$\frac{DT^{\alpha\beta}}{\partial x^{\beta}} = T^{\alpha\beta}_{;\beta} = 0 \tag{7.21}$$

what this equation implies is that the amount of energy E passing through the room you're in, per unit time is the energy current, $T^{0\alpha}$. For example, if $\alpha = 2$, then the flow of energy along $x^2 = y$ direction is equal to T^{02}, while T^{00} denotes the energy density. In the coordinate-independent sense, this yields a conservation law for the stress-energy tensor as: $\nabla \mathbf{T} = 0$.

7.5 And Finally, The Field Equations...

Let's place a source with mass (that curves spacetime) in the room you reside in, we know from the previous section that this geometric object's stress-energy tensor $\mathbf{T}$ must have zero divergence $\nabla \mathbf{T} = 0$, because of the conservation of energy-momentum. This source which has mass will communicate to the space around it and give it instructions on how to curve, i.e., it is responsible for the *generation* of gravity. Now, we know that when mass tells space how to curve, space in response will tell matter how to move. Thus, there exists a completely geometrical *object* which is proportional to the stress-energy tensor. This *object* must possess similar characteristics as $\mathbf{T}$, i.e., it must be symmetric and also have its own personalized conservation law (i.e., it must be divergence-free). This *object* happens to be the *Einstein tensor* $\mathbf{G}$ (because as mention in conservation, the conservation law $d^*T = 0$ is a consequence of the contracted Bianchi identity $d^*G = 0$). From the above reasoning, we can express the relation as follows

$$\mathbf{G} = \zeta \mathbf{T} \tag{7.22}$$

where ζ is the proportionality factor, which will be revealed later. Remember that we wanted to make a connection between the field and the source in such a way that the conservation of the source shall be an automatic consequence of some condition imposed on the field. The zero divergence of the Einstein tensor is the condition here and this automatically leads to the conservation of the stress-energy tensor. Now, if **G** is providing subtitles for the conversation between the geometry and mass, then it's language (i.e., the field's language) must be in terms of the metric tensor and curvature tensor. To see this relation and the conversation between mass and geometry in curved spacetime we are to first understand how the conversation plays out in flat spacetime, i.e., in Newton's version. We go back to the debate that Einstein and Newton had, and now Pauli questions them,

> How does matter tell space to curve? And how does space tell matter to move?

Newton says that since potential ϕ exists everywhere in space and since it varies from place to place, the field tells particles how to move by putting a limit on acceleration as follows

$$\begin{aligned} &\mathbf{F} = m\mathbf{a} = -m\nabla\phi\left(x\right) \\ &\mathbf{a} = -\nabla\phi\left(x\right) \quad \text{[Field informs particles how to move]} \end{aligned} \tag{7.23}$$

he further proceeds to tell that it is the *Poisson equation* that informs the gravitational field how to curve and concludes his thought.

$$\nabla^2\phi\left(x\right) = 4\pi G\rho\left(x\right) \quad \text{[Mass tells space how to curve via density]} \tag{7.24}$$

When its Einstein's turn he says that he would like to build on Newton's argument and he establishes that for a spherically symmetric source, the solution to the Poisson equation outside the gravitating source is as follows

$$\phi = -\frac{MG}{r} \tag{7.25}$$

he then proceeds to draw parallels between this finding and his friend, Karl Schwarzschild's formula that explains the geometry outside a gravitating source (i.e. the Schwarzschild metric) as follows

$$
\begin{aligned}
ds^2 &= \left(1 - \tfrac{2MG}{r}\right) dt^2 - \left(1 - \tfrac{2MG}{r}\right)^{-1} dr^2 - r^2 d\Omega^2 \\
g_{tt} &= g_{00} = \left(1 - \tfrac{2MG}{r}\right) = (1 + 2\phi)
\end{aligned}
\tag{7.26}
$$

Now that he has established a relation between the metric tensor and the Newtonian potential, he goes on to observe that there exists a Poisson-like equation in terms of the time component of the metric tensor as follows

$$\nabla^2 g_{00} = 2\nabla^2 \phi = 2\,(4\pi G\rho) = 8\pi G\rho \tag{7.27}$$

Einstein finally concludes that matter, via density, tells space how to curve by affecting the geometry (in terms of the metric). Well, Einstein is absolutely correct, but let us try and understand this Poisson-like equation that Einstein had built in his head. Einstein's generalization of the above equation first involves accounting for an *object* that is built out of derivatives of the metric tensor, which implies that it has to account for the geometry and not the matter; this turns out to be the Einstein tensor **G**. The other generalization involves accounting for all the components of the source, i.e., accounting for all the messengers (like density) of the source that communicate to space; this turns out to be the stress-energy tensor **T**. Thus, in component form, we have

$$G^{\mu\nu} = 8\pi G T^{\mu\nu} \tag{7.28}$$

our aim now is to arrive at a plausible representation for the Einstein tensor. Since **G** is proportional to the stress-energy tensor **T**, it must also possess properties of the same. Thus, a suitable candidate for the Einstein tensor must be a two-tensor, must be symmetric, and must possess derivatives of the metric tensor. The reason for the last condition is that the metric tensor by itself has the physical meaning of potential, and its derivative has the physical meaning of a field, similarly, the dynamical motion of the field can be described by the derivative of the field which is nothing but the second order derivative of the metric tensor. In short, we need a candidate which must represent the geometry of the field and be able to communicate to matter and inform it how to move. Now, we know that the Christoffel symbol possesses first order derivatives and has the form of $\Gamma \sim \frac{1}{2} g^{-1} \partial g$, but since it's not a tensor, it is not a suitable candidate. What about the Riemann tensor? Well, it takes up the form: $R \sim \partial\Gamma + \Gamma\Gamma - \ldots$, and since the Christoffel symbols contain the first order derivatives of the metric tensor, it automatically implies that the Riemann tensor possesses the second order derivatives of the metric tensor by taking up such a form: $R \sim \partial^2 g + (\partial g)^2 + \ldots$, and thus making it a suitable candidate. But wait! The Riemann tensor is not a two-tensor, it has four indices. Not to worry because it can always be contracted to obtain the Ricci tensor (which is a two-tensor) which takes up the following form

$$
\begin{aligned}
R_{\mu\nu} \equiv R^{\alpha}_{\mu\alpha\nu} &= \Gamma^{\alpha}_{\mu\nu,\alpha} - \Gamma^{\alpha}_{\mu\alpha,\nu} + \Gamma^{\alpha}_{\beta\alpha}\Gamma^{\beta}_{\mu\nu} - \Gamma^{\alpha}_{\beta\nu}\Gamma^{\beta}{}_{\mu\alpha} \\
&= g^{\delta\alpha} R_{\delta\mu\alpha\nu} \\
&= \tfrac{1}{2} g^{\delta\alpha} \left(g_{\delta\nu,\mu\alpha} + g_{\mu\alpha,\delta\nu} - g_{\delta\alpha,\mu\nu} - g_{\mu\nu,\delta\alpha} \right) \\
&+ g^{\delta\alpha} g_{\lambda\xi} \left(\Gamma^{\lambda}_{\mu\alpha}\Gamma^{\xi}_{\delta\nu} - \Gamma^{\lambda}_{\mu\nu}\Gamma^{\xi}_{\delta\alpha} \right)
\end{aligned}
\tag{7.29}
$$

From the above equation we clearly see that the Ricci tensor is composed of second order derivatives of the metric tensor hence making this the perfect candidate. It is important to note that the Ricci tensor by itself can be expressed in a further simplified form as $R^{\mu\nu} = g^{\mu\nu} R$, where R is called the *scalar curvature*. Enter the dilemma due to this simplification made–what candidate is suitable and which to choose, it is here that we resort to energy conservation. We are well aware of the conservation imposed of the stress-energy tensor (as a consequence of the contracted Bianchi identity remember?), and hence in the component form we write that the covariant derivative of the stress-energy tensor is zero, i.e., $D_\mu T^{\mu\nu} = 0$ (and just like $d^*T = 0$ followed from $d^*G = 0$, this equation follows from $D_\mu G^{\mu\nu} = 0$). Following from this, we have

$$
D_\mu \left(g^{\mu\nu} R \right) = g^{\mu\nu} \left(D_\mu R \right) + R \left(D_\mu g^{\mu\nu} \right) \tag{7.30}
$$

we know from the local flatness condition (i.e., in the local Minkowski reference frame where we applied Gaussian normal coordinates) that $D_\mu g^{\mu\nu} = 0$. Thus, the second term on the RHS vanishes leaving us with $D_\mu \left(g^{\mu\nu} R \right) = g^{\mu\nu} \left(D_\mu R \right)$. Oh, wait! I did mention to you previously that of all the two-rank tensors in the universe that we can form by contracting the Riemann tensor, it is only the Einstein tensor **G** that retains part of the Bianchi identities (and I also went on to mention the equation $G^{\mu\nu}_{;\nu} = 0$). Taking this as our hint, let's begin with the Bianchi identity and see if we can land up with a comfortable expression for $\mathbf{G}$. First, let's start by writing down the Bianchi identity for the Riemann tensor

$$D_\mu R_{\alpha\beta\chi\xi} + D_\xi R_{\alpha\beta\mu\chi} + D_\chi R_{\alpha\beta\xi\mu} = 0 \tag{7.31}$$

Upon multiplying by $g^{\nu\mu}g^{\alpha\chi}g^{\beta\xi}$ on either side of the equation (since the metric derivatives vanish, these act as constants and can be taken inside the derivative), we obtain the following

$$\begin{aligned} &D_\mu g^{\nu\mu}g^{\alpha\chi}g^{\beta\xi}R_{\alpha\beta\chi\xi} + D_\xi g^{\nu\mu}g^{\alpha\chi}g^{\beta\xi}R_{\alpha\beta\mu\chi} + D_\chi g^{\nu\mu}g^{\alpha\chi}g^{\beta\xi}R_{\alpha\beta\xi\mu} = 0, \\ &D_\mu g^{\nu\mu}R + D_\xi g^{\nu\mu}g^{\alpha\chi}g^{\beta\xi}R_{\alpha\beta\mu\chi} + D_\chi g^{\nu\mu}g^{\alpha\chi}g^{\beta\xi}R_{\alpha\beta\xi\mu} = 0, \end{aligned} \tag{7.32}$$

making use of the block symmetry $\left(R_{[abcd]} = 0\right)$ of the Riemann curvature tensor we express $R_{\alpha\beta\mu\chi} = R_{\mu\chi\alpha\beta}$, and $R_{\alpha\beta\xi\mu} = R_{\xi\mu\alpha\beta}$; hence we obtain

$$D_\mu g^{\nu\mu}R + D_\xi g^{\nu\mu}g^{\alpha\chi}g^{\beta\xi}R_{\mu\chi\alpha\beta} + D_\chi g^{\nu\mu}g^{\alpha\chi}g^{\beta\xi}R_{\xi\mu\alpha\beta} = 0, \tag{7.33}$$

making use of the first and second skew symmetries (i.e., $R_{abcd} = -R_{bacd}$, and $R_{abcd} = -R_{abdc}$), we express $R_{\mu\chi\alpha\beta} = -R_{\chi\mu\alpha\beta}$, and $R_{\xi\mu\alpha\beta} = -R_{\xi\mu\beta\alpha}$; hence we obtain

$$D_\mu g^{\nu\mu}R - D_\xi g^{\nu\mu}g^{\alpha\chi}g^{\beta\xi}R_{\chi\mu\alpha\beta} - D_\chi g^{\nu\mu}g^{\alpha\chi}g^{\beta\xi}R_{\xi\mu\beta\alpha} = 0, \tag{7.34}$$

using the Ricci tensor definition $\left(R^{ab} = g^{ac}g^{bd}R_{cd}\right)$, we write the following

$$\begin{aligned} &D_\mu g^{\nu\mu}R - D_\xi g^{\nu\mu}g^{\beta\xi}R_{\mu\beta} - D_\chi g^{\nu\mu}g^{\alpha\chi}R_{\mu\alpha} = 0, \\ &D_\mu g^{\nu\mu}R - D_\xi R^{\nu\xi} - D_\chi R^{\nu\chi} = 0, \end{aligned} \tag{7.35}$$

replacing dummy variables ξ, and χ with μ we have the following equation

$$\left[D_\mu g^{\nu\mu}R - D_\mu R^{\nu\mu} - D_\mu R^{\nu\mu}\right] = 0. \tag{7.36}$$

We can now factorize the derivative to obtain

$$\begin{aligned} D_\mu g^{\nu\mu} R - 2 D_\mu R^{\nu\mu} &= 0, \\ D_\mu \left(R^{\nu\mu} - \tfrac{1}{2} g^{\nu\mu} R \right) &= 0. \end{aligned} \tag{7.37}$$

On comparison of this result to the identity that we obtained via energy conservation we can conclude the following

$$\begin{aligned} & D_\mu R = \tfrac{1}{2} g^{\mu\nu} \partial_\mu R \overset{leads\ to}{\Longrightarrow} D_\mu R^{\mu\nu} = \tfrac{1}{2} D_\mu \left(g^{\mu\nu} R \right) \\ & \overset{which\ is}{\Longrightarrow} D_\mu \left(R^{\nu\mu} - \tfrac{1}{2} g^{\nu\mu} R \right) = 0 \end{aligned} \tag{7.38}$$

Observing the equation, we find out that this object possesses the second order derivatives of the metric tensor is a two-tensor (which is symmetric), and most importantly, retains a part of the Bianchi identity (this identity that is being conserved is called Bianchi's first identity). We also observe that the divergence of the tensor is null. Thus, this is the perfect candidate for our Einstein tensor $G^{\mu\nu} = \left(R^{\nu\mu} - \tfrac{1}{2} g^{\nu\mu} R \right)$. We can now write the Einstein's field equation in its complete form as follows

$$R^{\nu\mu} - \frac{1}{2} g^{\nu\mu} R \equiv G^{\mu\nu} = 8\pi T^{\mu\nu} \tag{7.39}$$

Remember the proportionality constant ζ, which related the stress-energy tensor $\mathbf{T}$ to the Einstein tensor $\mathbf{G}$? Now, in comparison, we can conclude that $\zeta = 8\pi$. What this field equation told Einstein was that the source of the gravitational field is not limited to energy density but also depends upon the flow of energy, the flow of momentum, and the momentum density. The components of the stress-energy tensor have the dimensions of energy density, i.e., $T^{\mu\nu} \sim ml^2 t^{-2} \sim mc^2 l^{-3}$; the Christoffel symbol has the dimensions of inverse length, i.e., $\Gamma^{\alpha}{}_{\mu\nu} \sim l^{-1}$; and the Ricci tensor has the dimensions of inverse square length, i.e., $R^{\mu\nu} \sim l^{-2}$. Thus, in order to maintain dimensional homogeneity, we require the constant κ to have the following dimensions: $\kappa \sim m^{-1} l c^{-2}$. This implies that the κ, in terms of the fundamental constants, has the following form: $\kappa \sim G/c^4$. Previously, we did not include this constant since we were working in natural units (i.e., $G \equiv c \equiv 1$). We can now rewrite the equation in terms of this constant (called *seminal Newton's constant*) as follows

$$R^{\mu\nu} - \frac{1}{2} g^{\mu\nu} R \equiv G^{\mu\nu} = \frac{8\pi G}{c^4} T^{\mu\nu} \tag{7.40}$$

The equation is also written with the additional term of λg, where Λ is called the *cosmological constant.* Einstein introduced the cosmological constant when looking for a stationary model for the cosmos and did not include it in he equation (which he later called *the greatest blunder of my life*). The equations with the cosmological constant included read

$$G^{\mu\nu} + \Lambda g^{\mu\nu} = \frac{8\pi G}{c^4} T^{\mu\nu} \tag{7.41}$$

7.6 Properties of the Einstein equations

Let's first analyse the fate of the equation in vacuum with no sources. This situation implies that $\Lambda = 0$ and $T_{\mu\nu} = 0$. Plugging these into the equation we get

$$R\mu\nu - \frac{1}{2} g_{\mu\nu} R = 0 \tag{7.42}$$

multiplying by $g^{\mu\nu}$ on either sides we get

$$\begin{aligned} R_{\mu\nu} g^{\mu\nu} - \tfrac{1}{2}\delta^{\mu\nu}_{\mu\nu} R &= 0 \\ R_{\mu\nu} g^{\mu\nu} = 2R &\overset{implies}{\Longrightarrow} R = 0 \end{aligned} \tag{7.43}$$

upon substituting this result in the vacuum, source-less version of the field equation we obtain

$$R_{\mu\nu} = 0, \tag{7.44}$$

this is called the *Ricci flatness condition.* It is important to note that the Ricci flatness condition does not imply vanishing curvature of spacetime, i.e., $R_{\mu\nu\alpha\beta} = 0$. There is something hidden in the field equation, it's trying to convey to us a key information. Let's see what this is. Take the covariant derivative of the equation with respect to the basis ν to get

$$D_\nu\left(R_{\mu\nu} - \tfrac{1}{2}g_{\mu\nu}R - \tfrac{1}{2}\Lambda g_{\mu\nu}\right) = D_\nu\left(\tfrac{8\pi G}{c^4}T_{\mu\nu}\right),$$

$$D_\nu\left(R_{\mu\nu}\right) - \tfrac{1}{2}g_{\mu\nu}D_\nu\left(R\right) - \tfrac{1}{2}RD_\nu\left(g_{\mu\nu}\right) - \tfrac{1}{2}\Lambda D_\nu\left(g_{\mu\nu}\right) = \tfrac{8\pi G}{c^4}D_\nu\left(T_{\mu\nu}\right), \tag{7.45}$$

notice that the terms containing $D_\nu\left(g_{\mu\nu}\right)$ would vanish due to the local flatness condition. We can now manipulate the leftover terms to get

$$g_{\nu\nu}\left(D_\nu R^\nu_\mu\right) - \frac{1}{2}\left(D_\mu R\right)g^\mu_\nu g_{\mu\nu} = \frac{8\pi G}{c^4}\left(D^\nu T_{\mu\nu}\right)g_{\nu\nu} \tag{7.46}$$

Dividing this equation by $g_{\nu\nu}$ on either sides we obtain the following

$$R^\nu{}_{\mu;\nu} - \frac{1}{2}\partial_\mu R = \frac{8\pi G}{c^4}D^\nu T_{\mu\nu} \tag{7.47}$$

observe the LHS, does it ring a bell? Yup, it's the first Bianchi identity, and since we already know that $R^\nu_{\mu;\nu} = \frac{1}{2}\partial_\mu R$, we can substitute this into the equation to obtain a very elegant result

$$D^\nu T_{\mu\nu} = 0 \tag{7.48}$$

We have just obtained back the conservation condition! This was the exact same condition we deduced in the earlier section. But wait, isn't something off here. We assumed that conservation holds and hence obtained the field equation, but what does it mean when the field equation itself reproduces back the same condition, what is the equation trying to tell us? The answer is that the conservation of the stress-energy tensor is just a consequence of the Einstein field equations.

References

1. Spivak, M. D. (1970). A comprehensive introduction to differential geometry. Publish or perish.
2. Fock, V. (2015). The theory of space, time and gravitation. Elsevier.
3. Mller, C. (1972). The theory of relativity.
4. Misner, C. W., Thorne, K. S., & Wheeler, J. A. (2017). Gravitation. Princeton University Press.
5. Weinberg, S. (1987). Anthropic bound on the cosmological constant. Physical Review Letters, 59(22), 2607.
6. Schutz, B. F. (1980). Geometrical methods of mathematical physics. Cambridge university press.
7. Isham, C. J. (1999). Modern differential geometry for physicists (Vol. 61). World Scientific Publishing Company.
8. Warner, F. W. (2013). Foundations of differentiable manifolds and Lie groups (Vol. 94). Springer Science & Business Media.
9. Sternberg, S. (1999). Lectures on differential geometry (Vol. 316). American Mathematical Soc..
10. Hawking, S. W., & Ellis, G. F. R. (1973). The large scale structure of space-time (Vol. 1). Cambridge university press.
11. Choquet-Bruhat, Y. (2009). General relativity and the Einstein equations. Oxford University Press.
12. Choquet-Bruhat, Y., de Witt, C., DeWitt, C. M., DeWitt-Morette, C., Dillard-Bleick, M., & Dillard-Bleick, M. (1982). Analysis, manifolds, and physics. Gulf Professional Publishing.
13. Nakahara, M. (2003). Geometry, topology and physics. CRC Press.
14. Faber, R. (2017). Differential geometry and relativity theory: an introduction. Routledge.

Index

www.ingramcontent.com/pod-product-compliance
Lightning Source LLC
Chambersburg PA
CBHW040318220526
45473CB00009B/2478
9781717858337